The PRINCIPLES of LANGUAGE

Towards trans-Chomskyan Linguistics
rewritten in 1919

Gero Jenner

Copyright © Gero Jenner, 2019

ISBN: 9781087335278
Imprint: Independently published
last revision 20.12.21

Dedicated to Otto Jespersen, Edward Sapir and John Lyons, those experts on language from whose soundness of theoretical thinking, openness of mind and practical inventiveness I learned most.

Preface _____ 1

Chomskyan and trans-Chomskyan Grammar _____ 5

    Chomsky's trees - the core of his theory of language _____ 6

Some further remarks on the evolution of language _____ 17

Meaning, Form and Formal Realization _____ 21

    Laws and Variety _____ 24

Some cautionary remarks _____ 25

    Difficulties arising from Meaning _____ 25
    Meaning - the Core of Language _____ 27
    Difference between linguistics and logic _____ 28
    Meaning and its Realization in Form _____ 29
    Old difficulties removed _____ 30
    Explanatory and Descriptive Grammars _____ 31

Analytic and Constructive Linguistics _____ 35

    1 Explaining Generativeness _____ 36

    2 Concerning method _____ 37

    3 Concerning the use of traditional terms _____ 38

    4 Formal relevance _____ 38

    5 Synthetic, Agglutinative, Fusional, Polysynthetic Languages __ 39

I Basic Terms of Universal Grammar _____ 41

II The Logical Structure of Meaning _____ 45

    1 Semants – the Subunits of Meaning _____ 45

    2 Syntheses - the Units of Meaning _____ 46

    3 Quantitatively Enlarged Synthesis - Ranks _____ 47

        More instances of the Enlarged Synthesis _____ 49
        Semantically effaced Enlargements _____ 51
        Psychic-state Syntheses _____ 51

    4 Semantic inversion _____ 53

5 Connections and Conjunctions _____ 56
Different Formal Realizations of semantically identical conjunctions _____ 58

III The Informational Structure of Meaning _____ 61

1 Commands (Requests) _____ 62

2 Statements and Questions _____ 62

3 Topic versus Novum _____ 64

4 Free versus bound synthesis _____ 69
Categories and the bound synthesis _____ 71
Traditional concepts in Comparative Linguistics _____ 72
Every synthesis may be either free or bound _____ 74
General pattern of formally realizing the bound synthesis _____ 76
Not all languages formally realize the bound synthesis _____ 84
Practical definition of free versus bound synthesis _____ 87
No informational shifting within the bound synthesis _____ 88

5 Semantic effacement _____ 89
Total semantic effacement and rank-lifting _____ 94
Derivative use of total effacement _____ 96
Remarks on form: the so-called passive voice in traditional grammar _____ 98

6 Semantic Tingeing _____ 99

7 Appendix: the Frozen Synthesis - the genesis of concepts _____ 100

IV The Formal (Symbolic) Realization of Meaning _____ 105

1 Formal means in natural language _____ 105

2 Phonetics _____ 106

3 The Differentiation-Value _____ 106

4 Formal Equivalence, Deficiency, Abundance _____ 107

5 Formal Extension _____ 109

6 Morphology _____ 112
Differential Analysis _____ 114

V Syntax and Paratax - Basic Modes of Formal Realization _____ 119

1 Shortcomings of Chomsky's Generative Grammar _____ 120

2 Usefulness of traditional terminology _____ 121

VI Law in Language _____ *123*
   1 First basic law (concerning formal equivalence) _____ 124
   2 Second basic law (formal realization of open field semantic categories) 126
   3 Third basic law (the use of position in natural languages) _____ 127
   4 Fourth basic law (Pre- or postpositions) _____ 130
   5 Fifth basic law (concerning morphology) _____ 131

Index _____ *133*

Preface

Ferdinand de Saussure studied the relationship between désignant (form at the level of word units) and désigné (meaning at the level of word units). He found that it is arbitrary in human languages, since any word is capable of expressing any meaning. De Saussure did not explore the more fundamental question of the relationship between generalized form (the structured sound waves used by human speakers in what we call language or communication) and generalized meaning (the set of structured meaning expressed in human languages). But this is the first fundamental question that any general theory of language must answer. Is this relationship arbitrary as well? Definitely not! It is governed by both arbitrariness **and** laws.

The second and third basic questions concern the two constituents themselves, that is generalized form (= désignant beyond the level of word units) and generalized meaning (= désigné beyond the level of word units), which are taken as given by de Saussure and most linguists. The generalized form of human speech (i.e., structured sound waves) is conditioned both by the properties of the human speech organ and by human memory. As for generalized meaning, Aristotle and Port Royal had already tried to open this field of research.

These are the three basic questions to be answered by general linguistics. It should start from de Saussure and from the insights of Aristotle and Port Royal. Which means: if we want to get beyond Chomsky, we must go back behind Chomsky.

I consider Pure Meaning as the basis of language – meaning, which becomes embodied in form, that is, in structured sequences of acoustic waves. It should be noted that structured vibrations of the air, that is form, fundamentally differ from meaning in a very precise understanding: there is no way *of deriving any meaning from mere vibrations of the air* or from their representation as graphic symbols (letters, words, sentences) on a sheet of paper. Acoustic waves or their graphic representations are merely *assigned* to meaning so as to evoke it. It was the basic error of Chomsky's so-called Generative Grammar to have missed this essential point.

On the other hand, it is perfectly possible to construct machines that totally ignore meaning. They transform the vibrations of air belonging to some language A into acoustic waves correlated with a language B (or do the same for their respective graphic representations). In the case of basic items like words this transformation is laid down in dictionaries where German 'Baum' is identified with English 'tree', without the dictionary being aware of any meaning. In the case of larger formal units like sentences, translation machines proceed in a similar manner. In many cases, the transformations of larger units like sentences do, however, lead to spurious results. In order to exclude such errors, reference to the linguistic environment is mostly sufficient. This means that a *broader* formal environment determines a *narrower* one, so that such reference mostly produces appropriate translations. As the broader environment still exclusively consists of other formal elements (acoustic waves or their graphic representations), *translation machines may rely exclusively on form* and be highly reliable (which is certainly true of the most developed among them).

Such basic formal procedure was initiated by Distributional Analysis, which has thus made an immense contribution to the pragmatic purpose of translating languages. But translation machines do not further our *understanding of language* - understanding is indeed quite a different matter. As Distributional Analysis could keep meaning strictly out of its way, it has on the contrary be an obstacle to the understanding of language. The reason should be perfectly clear. *Meaning represents the very fundament of language.* It is of primary importance when we try to explain man's generative linguistic capacity since there is nothing in form (mere vibrations of the air) from which to derive it.

Chomsky must have felt the shortcomings of his theory, as he tried to amplify it by means of a concept he called *'deep structure'* - opposed to a so-called *surface*. If the first was to have any sense at all, it should refer to what lies at the bottom of form, namely meaning ("Mentalese" as termed by Steven Pinker). But meaning could not be arrived at by purely formal Distributional Analysis, so Chomsky got stuck and soon abandoned Linguistics altogether – it so happened that after him General Linguistics got stuck as well.

The impasse was obvious, for Chomsky was right and wrong at the same time. Meaning was of no use to the soon flourishing new science of computerized translation, so many linguists felt sure that they need not bother about it. As for Chomsky himself, his avowed and ultimate aim was to explain man's linguistic generativeness – and he was totally wrong when he believed that this goal could be reached without reference to what is the very core of language. In order to lead to a true theory of generativeness, the deep-structure would have to embody what I call 'Pure Meaning', while the surface-structure would be pure form, that is sequences of acoustic waves (or their graphic representations) described by means of Distributional Analysis.

Let me illustrate this basic point right at the beginning:

First, historically (or phylogenetically): Animals already conceive reality and act accordingly, even if they do not translate these conceptions into auditory or any other signs and signals.

Second, ontogenetically: Infants add auditory signals ('signifiants' or words) only to concepts already present in their mind, otherwise they would pronounce empty sounds. Mere crying is, of course, meaningful too as it usually is the outward expression of pain.

Third, pathologically or ‚a contrario': Deaf-mute persons (like Helen Keller) are equipped with ‚meaning' independently of its formal realization in sound structures. The fact is proven by their ability to replace sounds with a sign language consisting of gestures.

Fourth, pragmatically through translation: When translating an English sentence into Chinese, I must, first, go back to its meaning before, subsequently, applying the specific rules governing its formal realization in Chinese (as mentioned above, computerized translation-machines proceed without reference to meaning because the reference is intuitively carried out by human programmers relying on meaning: tree = arbre because of meaning).

And *fifth*, methodically: The preceding considerations acquire their most general significance as soon as we switch to comparative linguistics. There are but two *tertia comparationis* between any too randomly chosen languages. These lie at the bottom of any specific translation as well as of any general statement about comparative linguistics and linguistic laws. The first tertium comparationis consists in *Pure Meaning* (that is meaning apart from and prior to any realization in form by sounds, letters, gestures etc.). In other words, it is what Steven Pinker has called 'Mentalese'. The second tertium comparationis are the *Formal Means* at the disposition of human beings.

Generativeness

What such an approach aims at should be evident: It endeavors to explain, first, the *particular* generativeness of the single speaker of some given language, say English. What is it that enables him to create an *infinite set of sentences* after having been acquainted with only a finite number of such – mostly when being a child? Second, this approach is meant to explain *general* generativeness. What are the necessary conditions and the constraints that govern the human capacity for creating *any natural language whatsoever*, that is those which have already been created in the past and those new ones he may still create in the future?

Somebody has termed my approach 'pragmatic' – a misnomer. Computerized translations that transform a formal sequence of some language A into the formal sequence of some language B deserve such a characterization as they may be put to pragmatic use. The approach expounded in this book is not pragmatically useful in this sense. If it has any merits at all, it is its capacity to make us *understand* the true nature of meaning and its relation to form in natural languages. Noam Chomsky had conceived the grandiose idea that a speaker must dispose of a set of rules by means of which the science of language may explain his particular generativeness. Chomsky even hinted to general generativeness when referring to innate ideas at the base of human linguistic capacity. But Chomsky was unable to prove his point. Generative grammar based on Distributionalism, that is on a purely formal procedure (like machine translation) is by its very method barred from proving what Chomsky wanted to prove.

Preface

Chomskyan and trans-Chomskyan Grammar

Those interested in the logic of language will be rewarded by reading this book, for it reveals and explains the boundary between linguistic chance and linguistic law, which exists both in language as in culture in general, but is much easier to determine in language. As a matter of principle, immaterial conceptual meaning and its material manifestation through sound sequences that are exchanged in the process of communication between speaker and listener, are regarded as the two constituent components of language and carefully kept apart.

The conclusion of the present book proves Chomsky right: Yes, there is a General and Generative Grammar. Language *is certainly generative* because children are capable of forming an infinite number of statements, even if they have never heard them before. And, yes, the faculty of language *must be general* because the statements of different languages can be translated into each other. These are empirical facts. But language is not generative and general according to the deceptive simplicity of the model illustrated by Chomsky when he presented those (once famous) inverted trees. *At the top of the tree he wrote an S for sentence, from which a speaker was supposed to derive in downward direction all possible concrete instances of that language with the help of but a very few general rules and a lexicon. Each particular language then adds some specific rules to the general ones in order to define the differences with respect to other languages.*[1] That was the dazzling idea of the Chomskyan model, its actual core, while everything else was just ancillary. The model owed its fascination to the fact that it turned language into a kind of rather simple computer game.

But language is not that simple,
this model is wrong from the outset, because its basic concepts (S, NP, VP, V, N etc.) are hybrid - they mix up the deep level of the immaterial analysis

---

[1] For example, the difference in word order, which in English mostly prescribes a middle position of the verb, i.e. SVO, whereas in Japanese it prescribes its position at the end: SOV.

of reality (the conceptual structure) and its material manifestation by means of acoustic (or other) signs. Immaterial reality analysis already takes place in animals even without the use of material signs, and it develops in humans from primitive beginnings (as in the Amazonian Piranha language, for example) to the most complex conceptual structures. These, however, are based on a broad general structure only - while sentences from an evolutionary primitive language can be easily translated into a more developed one, this is very difficult or even impossible in the opposite direction (how can a modern text on mathematics be translated into a language where people don't use numbers beyond two or three?).

But differences on the conceptual level do by no means exhaust the complexity of language, because on the basis of identical immaterial conceptual structures various material realizations, i.e. sign systems, can be built. Chomsky's seductively simple tree does violence to language and it explains strictly nothing. In the present book, General and Generative Grammar is presented as a complex ensemble, which furthermore is characterized by constant evolutionary unfolding.

Chomsky's trees - the core of his theory of language

The fascination of Chomsky's theory of language is due to the fact that it seems to derive linguistic diversity and complexity from a simple starting point. After Chomsky, a whole generation of linguists was busy with drawing these elusive branched trees. Let us stick to a simple example:

```
                          S
            NP                          VP
      det        N         V        det        N
      The       boy       eats       the     ice cream
```

The derivation is fascinating because of its apparent proximity to the approach of the natural sciences, where complex events are similarly derived from simple basic elements. No wonder that many praised Noam Chomsky's approach as a revolution that finally turned the study of language into a science. The tree, with its simple peak of "S" for S(entence), seemed to define the rules that a speaker must obey in order to "generate" a potentially infinite

Preface

number of grammatically correct sentences in English or any other language (hence the name "Generative Grammar").

But right at the beginning linguists
should have asked the crucial question, what "S" at the top of the derivation is meant to represent? "S" cannot be an entity void of any content, as something cannot be derived from nothing. It must have some definite content. But what exactly?

As a matter of course, "S" cannot be identical with the formal end product, i.e. the English sound sequence "The boy eats the ice cream", because then there would be no derivation at all, but the whole thing would amount to *a mere tautology*. Nor can "S" be a composite of meaning and form, in the way the English word "boy" represents a phonetic form on the one hand and a carrier of meaning on the other. Then we would end up with a *partial tautology* since the formal end product is derived from a similar formal input.

The only possible interpretation is
that "S" at the top refers to something quite different: a structure of pure meaning not yet transformed into a sequence of sounds (or its graphic written counterpart). In the speaker's brain, the real event is present in a purely conceptual shape, which in the act of speaking he translates into a structured linguistic form.

But then "S" as a term for S(entence) or formal structure turns out to be a misnomer. We have to replace it with another expression, say "M" as an abbreviation for M(eaning):

```
                         M
          **********************
       NP                            VP
   det      N                    V          det      N
   The      boy                  eats       the      ice cream
```

However, once we perform this necessary reinterpretation, it becomes obvious that we must separate the starting point "M" with a line from what follows, because NP and VP represent something quite different from meaning, *namely formal elements in a given temporal order*. "The boy" (NP) proceeds in temporal sequence (VP) that is "eats ice-cream". Such temporal order is not found in the conceptual structure itself. A unit of meaning such as "The tree is green" is independent of time. And this equally applies to a unit of meaning like "The boy eats ice cream". When happening in outside reality, the action of eating is of course a temporal event like any other action, but it has nothing in common with the sequence of words in the English sentence.

And "M", which we have to substitute for "S"
exhibits still one more distinguishing feature. The expression "S" suggests unity and simplicity, which, however, does not at all exist on the level of meaning. "The tree is green" denotes the modification of a substance by a quality. "The boy eats" or "The boy eats ice cream" refers to the modification of a person (living substance) by an action. The "Logical structure of Meaning" (see my work *"The Principles of Language - Towards trans-Chomskyan Linguistics"* portrays the most important types of such units of meaning. Each of these can take the place of "M" on the top of the tree.

Since, furthermore the conceptual analysis of reality begins in the animal kingdom and is subject to evolution in human societies as well, only the basic types are present in all societies, not their more complex forms. In other words, evolution already comes into play at the level of "M".

Dealing with Meaning involves evolution
The Logical Structure of Meaning comprises the elementary types of synthesis like "Paul dances", "Trees are green", "It happens now". "He lost it here" etc. Even animals must have analyzed reality according to whether something represents an action in time or a more or less constant property (quality) and whether it happens here or there, now or in the past, etc. Even if they don't use auditory or other signs, that is a language, in order to transmit such analysis to their fellows, their brains must be capable of performing such operations, otherwise they would not be able to adapt to an outward world characterized by unchanging properties as well as changing events.

But even chimpanzees trained in using different chips as substitutes for auditory signals only transmit claims (demands) or warnings like "I want to have a banana" or "Beware of such and such predator!" *Animals don't utter statements or questions*. In other words, the "Informational Structure of Meaning" not only differs from its logical counterpart in so far as it describes how the material provided by the latter is being put to the service of information between individuals or in societies, but it exhibits one more very

Preface

distinctive feature. *The "Informational Structure of Meaning" is a purely human achievement.* But it too is subject to evolution - Homo Sapiens did not create it ready made all at once. While the dichotomy of question versus statement is to be found in the most primitive languages recorded, that of bound versus free synthesis is a product of later evolution as is true of rank lifting and other more subtle needs of information. And the evolution of informational needs may not have come to an end at the present stage.

What about the components N and V of the formal level
below "M", that is below the structure of meaning? According to Chomsky, these belong to General Grammar, so that we may apply them to languages as different from each other as English and Chinese. But are these terms in fact universal? No, they are definitely not. Here again flawed logic unites with lacking empirical knowledge, when Chomsky asserts that they are.

Supposed that in all languages "verb" represents a formal slot (paratactic class) exclusively filled with the semantic category of actions (run, eat, take, play etc.) and "noun" a formal slot exclusively filled with living or non-living substances (house, cloud, tree, tiger etc.), then we would indeed have universal categories as we may be sure to find actions and substances in all natural languages. But this definition is contradicted by linguistic reality. In English, words such as "running", "speaking", "striking", etc., formally belong to the class of nouns although they express actions.

The conclusion therefore seems evident: it is impossible to define verb or noun in a general (universally valid) way. All we know from empirical data is that *different languages create their own specific formal classes* comprising actions, substances, qualities etc. Again, we have to modify Chomsky's deceptively simple scheme:

$$M$$
$$*********************$$

NP$_{eng}$		VP$_{eng}$		
det$_{eng}$	N$_{eng}$	V$_{eng}$	det$_{eng}$	N$_{eng}$
The	boy	eats	the	ice cream
	Running	tops		walking

Preface

Now, consider another example to better understand this basic modification. In English we may say "(In my view) running tops walking", which we understand in the sense that someone prefers to run rather than just go walking. In many languages this content cannot be expressed in a similar way that is without formally omitting the reference to a specific agent. In some languages, people must, for example, say, "I like to walk, but I'd rather run." The agent "I" cannot simply be effaced like in English.

To sum up, Chomsky's scheme does not in any way describe the generative linguistic capacity of human brains. On the one hand, Chomsky's "S"(entence) is either tautological or has to be replaced by "M"(eaning) - and then becomes much more complex, since "M" consists of different conceptual types (described in the Logical and Informational Structures of Meaning). This is an error of logic. On the other hand,

We hit upon an empirical error - categories such as V and N
are not universal; when used as such, they obscure the existing differences in the formal realization of languages instead of explaining them. The error of mistaken universality is due to the fact that transforming structures of meaning into structures of sound does not only result in *differences of syntax*, i.e. in different temporal sequences (like SVO in English, SOV in Japanese), but creates *differences in paratax* as well. These concern the classification of semantic concepts in formal slots (paratactic classes) like English verbs, Japanese verbs, etc.

With their deceitful simplicity Chomsky's trees - the essence of what is methodically new in his linguistic theory - all but obscure our understanding of language. But the question why Chomsky created a scheme that so blatantly disregards basic logic and empirical knowledge, need not concern us here, I will discuss it at the end of the chapter.

Chomsky's simplistic trees need still one further correction
Unless they be tautological, all the expressions above the dividing line must refer to meaning, ie the immaterial conceptual structure, while all expressions below belong to the acoustic chain or its representation on a sheet of paper. Now, there is no cogent reason why in sentences like "The boy eats ice cream" or "running tops walking" the verbal phrase VP should be represented by "eats the ice cream" or "tops walking" rather than by "The boy eats" or "running tops". There is no justification for such classification

neither on the formal level below the punctuated line nor on the conceptual level above it. We will see later that there is such justification (on both levels) only in cases like "dirty cloth" or "chanting joyfully". So, we again modify Chomsky's tree leaving out NP and VP altogether:

```
                        M
        ***********************
det_eng      N_eng       V_eng    det_eng     N_eng
The          boy         eats     the         ice cream
             Running     tops                 walking
```

After this final transformation, Chomsky's modified and reduced tree corresponds exactly to the general formula I had already used back in the eighties:

$$M \ldots \textbf{transformed into} \quad F$$

where M refers to meaning and F to its transformation in symbolic form.

With this correction in mind, lets go back to our original example: "The boy eats the ice cream". It represents a conceptual structure consisting of Agent and Patient together with an Action. *I separate these members by commas in order to indicate that on the conceptual level there is no temporal sequence.* According to the specific rules governing English syntax and paratax, the conceptual structure is then transformed into the following acoustic chain or sentence:

Ag, Pt, a(ction)   **transformed into**$_{eng}$ The boy eats the ice cream

Or, if you prefer the shape of a tree: where "etc."

```
                        Ag, Pt, a
        ***********************
det_eng      N_eng              V_eng        det_eng     N_eng
The          boy                eats         the         ice cream
A            dog                devours      a           bone
etc.         etc.               etc.         etc.        etc.
```

11

Preface

where "etc." comprises the entire formal slot or paratactic class.

Both schemes distinguish in a perfect and unequivocal way a deep from a surface structure - the first representing pure conceptual meaning the latter its formal representation.

Chomsky inherited his approach and method
from his teacher Zellig S. Harris, the founder of distributionalism. Strictly excluding the semantic dimension, Harris had restricted the description of language to the study of recurrent formal elements. Let us consider the following utterance:

Birds are chanting joyfully: N   V   Adv
Mary washes all dirty cloth: N   V   det   Adj   N
Big clouds cover the sky: Adj N   V   det   N

Knowing that he may replace any noun, like for instance cloth, with a larger expression like dirty cloth, or any verb, like for instance chant, with a larger expression like chanting joyfully, the distributionalist may then write NP for N and VP for V:

Little Mary   eagerly washes   all dirty cloth
   etc.   etc.   etc.
   NP   VP   NP
               S

Any purely formal distributional analysis may, of course, be turned upside down. Then "S" is placed at the top but that won't change its nature: it remains strictly tautological, with "S" being a mere abstraction representing no more and no less than the respective formal chains in English.

    This is not the way chosen by Chomsky. *By a mere sleight of hand, Chomsky turned an analytical process - a tautology - into a derivation.* As mentioned before, the apparent miracle was nothing more than a logical error.

    It should however be noted that Distributionalism with its purely formal analysis of language, ie its complete avoidance of meaning, prepared the way for and the great success of machine translation. Machines must do without

Preface

meaning, while General and Generative Grammar are based on meaning - therein lies their basic difference.

The preceding arguments represent in a nutshell
my refutation of Chomsky's method when applied to General and Generative Grammar. The following remarks are meant to explain some further particulars of my procedure.

Let us consider the following spoken sequence - a rather simple one for that matter:

He walks extremely fast. Yes, walking is such a pleasure. To be sure!

Chomskyan grammar assigns the symbol S = sentence to all three expressions. But in form as well as in meaning they are totally different. Only the first represents a complete semantic expression, certainly to be found in all known languages. A living substance (man) is characterized by an action that is itself further specified.

The second sentence expresses the feeling of the person addressed regarding the event. But only developed languages permit the formal raising of an action (to walk) to the rank of noun (walking) with the concomitant effacement of the agent (the walking of any person whatsoever). Likewise the second noun "pleasure" is in the same way arrived at by rank-lifting with agent effacement. In more primitive languages this sentence would have to be expressed in a sentence like for instance "Yes, many people like to walk."

Such a basic example proves that the formal term "noun" cannot be used in General Grammar as its semantic content is different for different languages. The third sentence represents an affirmation, which could be expressed as an entire sentence as well, for instance: "I totally agree with you."

Starting from Harris and Distributionalism, Chomsky endeavored to find something like a universal algorithm which would allow to deduce from the apex of a tree defined in purely formal terms (S) all possible alternatives at its bottom by gradually stepping down. A simple glance at any examples like the above does, however, suffice to make this an impossible undertaking. Different languages just embody meaning in different ways so even the terms "noun", "verb" etc. do not represent universals.

Preface

On the other hand, Chomsky's procedure of purely formal analysis turned out to be extremely successful in an altogether different domain far away from universal grammar and not taken into account by Chomsky himself, namely in machine translation. Machines do not understand meaning, they can only handle purely formal terms. After being taught by some human being that English "tree" is equal in meaning to German "Baum", they have no difficulty in substituting one for the other, and after being shown that the English question "does he come?" must in German be rendered by a change in word order "Kommt er?", they easily apply this algorithm to all similar instances. Machine translation consists in replacing the formal pattern of some language A with the formal pattern of some language B - provided that the human mind first established their common meaning.

But algorithms must in some instances give way to specific substitutions that cannot be derived from any rule. The German suffix for past tense "-te" (lach-te, speis-te, freu-te etc.) may in most cases be replaced by the English suffix "-ed" but not in all (jump-ed, laugh-ed, but went, dealt, hit). Likewise there are lots of idiomatic expressions in English, German and any other language that must be dealt with as exceptions since they cannot be derived from any general rule.

Recursion, embedding
Both terms are of special importance within Chomskyan Grammar. It should however be noted that they are defined in a purely formal manner - if at all.[2] Something - whatever that may be - does either recur or become embedded within some larger formal entity. But on a purely formal basis repetitions like: "Dogs, cats, man etc. ran, jumped, stumbled to the promised land, the

---

[2] It is remarkable that two staunch defenders of Chomsky, Steven Pinker and J. Mendivil-Giro, interpret recursion as found in their master's theory in opposite ways. Mendivil-Giro ("Is Universal Grammar ready for retirement?"): "The mathematical concept of recursion was quasi-synonymous with computability, so that recursive was considered equivalent to computable... what Chomsky... postulates as the central characteristic of human language is recursion in the computational sense, *not the existence of sentences within sentences or the existence of noun phrases inside noun phrases*" (my italics). Pinker ("The Language Instinct"): "Recall that all you need for recursion is an ability *to embed a noun phrase inside another noun phrase or a clause within a clause*" (*my italics*). Is there a better proof of Chomskyan vagueness than such opposing interpretations?

paradise, and their destiny in an endless, densely packed queue" are part of the same linguistic device. That is, recursion and embedding must be analyzed according to their meaning if we want to get an insight into their possible range and importance.

When seen in this perspective, they are of primary importance in General Grammar as seen by the two following rather basic examples:

a) The green tree is small
b) The green tree is green

The second example provides an instance of recursion as a substance and its qualification (green tree) occur in two different positions as an independent and a dependent clause. Example b) is logically redundant because both the main and the subordinate clause convey the same information. It is therefore meaning that provides the *interlingual* yardstick by which to distinguish both. "Green tree" is a non-informational (bound) synthesis while "tree is green" represents its (free) informational counterpart - a basic distinction in Universal Grammar (see: chapter III: The Informational Structure of Meaning).[3]

The uselessness of traditional as well as Chomskyan concepts when describing the 'logical' and the 'informational' structures of meaning

Describing what so far has not been described, namely pure meaning (Mentalese in Pinker's terminology), will put the reader's patience to a test of endurance as most usual terms of grammar like noun, verb, adjective, subject, object, active versus passive voice etc. have no place either in the structure of Pure Meaning nor in that of Pure Form.

As I will show later, these terms are defined ambiguously by means of *both semantic and formal criteria*. While it is perfectly legitimate to speak of English, Japanese, Chinese nouns, verbs, adjectives, nominal phrases etc. (since they all contain substances, actions, qualities etc.), it is empirically wrong to suppose their semantic contents to be identical in these different

---

[3] Some further considerations in:
http://www.gerojenner.com/wpe/the-hallpike-paper-universal-and-generative-grammar-a-trend-setting-idea-or-a-mental-straitjacket/

Preface

languages. Instead of explaining linguistic variety these terms *explain it away*.

For this reason, the traditional concepts still used by Chomsky cannot be accepted as primary units in General and Generative Linguistics. It is not because of any personal leaning for neologisms that I use new basic terms when describing the Logical and the Informational Structure of linguistic meaning, but out of necessity – nobody has done so before. Steven Pinker was well aware that what he calls "Mentalese" - in other words pure meaning - constitutes the very basis of language, but he didn't want to further pursue this line or did not have the courage to do so.

Therefore, I had to start from scratch when developing the new paradigm. The two first attempts were "Grammatica Nova" (Peter Lang, 1981) and "Prolegomena zur Generellen Grammatik" (Peter Lang, 1991), both written in German. Finally, I published a more substantial work "Principles of Language" (Peter Lang, 1993). Though written in English, it was all but ignored - and for good reason. Like many beginners, I overburdened my demonstrations with technical abbreviations which instead of furthering understanding transformed it into an almost impossible task. Thirty years later - and perhaps a little bit more considerate - I am fully aware of the mistakes made at the time. However, these shortcomings do not concern the basic ideas - these have withstood the test of time and seem even more relevant today. If we want to understand what Steven Pinker had called Mentalese; or, more generally speaking, if we want to understand Linguistic Variety and Linguistic Law, in other words, if we want to establish Comparative Linguistics as a Science, then there can be no other paradigm than a trans-Chomskyan one. Let me add that in the present work I renounced all references, most footnotes and even bibliography - presenting only the new paradigm itself. Sources and references may still be found in the book printed thirty years ago.[4]

As I deal with the basics of Meaning, Form and Formal realization not properly understood even now, I have chosen the most simple and sometimes trite examples without regard to any literary amenity - even repeating them when it was expedient to elucidate further aspects. I hope the gain in clarity will make good for the loss in pleasure.

---

[4] Or in: http://www.gerojenner.com/mfilesm/OldPrinciplesofLanguage.pdf

Preface

Some further remarks on the evolution of language

Steven Pinker revealed - unintentionally, of course - the fundamental weakness of Chomsky's theory when introducing the amazing concept of Mentalese. The *pre-symbolic deep layer* of language in the human brain evoked by this term is indeed the true foundation which the *symbolic superstructure* is subsequently built upon. Every child does, of course, enter life with this equipment. Pinker is quite aware of the fact when he states: "there is a pre-established harmony between the mind of the child and the texture of reality (157); the child somehow has the concepts available before experience with language and is basically learning labels for them."

Indeed, this basic insight was already found and stated by Otto Jespersen in his "Philosophy of Grammar" one hundred years ago(1925:55): "We are thus led to recognize that beside, or above, or behind the syntactic categories which depend on the structure of each language as it is actually found, there are some extralinguistic /rather pre-symbolic/ categories which are independent of the more or less accidental facts of existing languages; they are universal in so far as they are applicable to all languages."

I will show that categories like action, substances, qualities do indeed constitute the very base of natural languages - and, what is as important, that they only allow for a quite limited number of combinations. For instance, an action like running cannot be high or yellow, such a combination can only apply to substances like trees etc.

The moment we accept such a basic structuring of reality by the human mind - we may call it a "deep structure of meaning" - the problem of evolution arises in a new way. Every language, even the most primitive ones described by Christopher R. Hallpike (*"So all languages aren't equally complex after all"*, 2018), differentiate between the basic categories just mentioned, but they proceed quite differently with regard to concrete semantic differentiation. Some distinguish a wide range of colors, smells or tastes, others do not. The development of the lexicon towards ever greater complexity obviously presupposes a corresponding social and technological development. Different social classes therefore often evolve their own more or less developed vocabulary. And script, the prerogative of the learned in earlier times and up to a certain measure even today, remains the most effective catalyst in furthering complexity (see Hallpike, op. cit.).

Nor is the deep structure of meaning limited to the *logical analysis* of reality. Equally important is the final goal of all languages: the transmission and exchange of information, i.e. the *analysis of information.* A sentence like English: "The man we saw yesterday on the hill will visit us today", appears in some languages (also described by Hallpike) in a very cumbersome way like: "You know, there is a man. We saw him yesterday on the hill. That man will visit us today." In English, the fact that the person being addressed already possesses a certain information is expressed by an appropriate syntactic construction serving just this purpose. Without such a ready-made syntactic form, a laborious repetition of information already known to the person addressed must be used.

What now is the relationship
between increasing semantic differentiation, translation of the semantic depth structure into symbolic surface structures, and possible advantages of adaptation? That there must be quite a close such relationship seems hard to deny, as long as we speak about representing the logical and informational depth structure in symbolic form (that is, as an ordered structure of sounds, gestures, writing etc.). There can be no doubt that this evolutionary "invention" has immensely facilitated and enhanced the transmission of knowledge between individuals and generations. For this reason, many animal species also communicate by sounds or signs, at least in rudimentary ways (dolphins, meerkats, etc.). On the pre-symbolic level, a similar advantage results from the expansion of the lexical basis. Even if we must renounce to assign an index to *individual* linguistic achievements in terms of adaptation advantages, I see no difficulty in explaining the evolution of language if we start from a deep structure of meaning and its symbolic representation on the level of form.

Within the framework of linguistic theory as here outlined,
the evolution of language does not present an unsurmountable challenge as it does for Chomsky. The separate consideration of pre-symbolic analysis of reality on the one hand and its symbolic (formal) realization through a system of signs on the other allows the genesis of language to go back to the

animal kingdom. Indeed, the pre-symbolic analysis of reality is already achieved in all higher developed animals. These distinguish not only the most diverse qualities (smells, colors, etc.) but also movable from immovable things, etc. - otherwise they would not be able to survive. And not a few animals have developed some language of signs (sounds, gestures, optical signals etc.) or they can be made to distinguish between dozens of symbols, like chimpanzees for example. Pinker's and Chomsky's understanding of language as a symbol-manipulating machinery (a computational mechanism) - all made of one piece so to speak-, where meaning has no place even in the deep structure, requires a supermutation, some deus ex machina, to be possible at all. In other words: *it is an obstacle to the progress of scientific understanding*. In the theory presented here, the evolution of language appears as a gradual process with no mysterious mutations required to explain it.

Unfortunately, most linguists who have studied universal grammar over the past half century have been less eager to understand language (let alone languages in the plural) than to understand Chomsky - a gigantic task, as even Pinker had to admit when he complained about his teacher's scholasticism. But the challenge of understanding the great man (because the concept of a Universal Grammar is a great idea!) has inspired a whole generation of linguists, and now, as the obvious weaknesses and contradictions of Chomsky's theory become more and more obvious, another generation feels inspired to unravel the tangle of Chomsky's scholasticism with equal zeal - which is at least as difficult. Yet the most important thing for linguists should be to simply get back to a better understanding of language and not of Chomsky! Which means, first of all, that they would be well advised to know more than just their own mother tongue!

This brings me finally
to the *symbolic representation* of the logico-informational basis. The linguist Nicholas Allott writes: "The simplest illustration of a parameter is the choice of Head first or Head last; depending on which choice is made, a language is either SOV or SVO with many associated orderings in other aspects of syntax... All they /children/ have to learn is whether their particular language has the parameter head-first, as in English, or head-last, as in Japanese." Since I lived in Japan for more than three years, I am quite familiar with the language. The method followed in the present work will enable us to prove what with Pinker remained a mere question mark, namely why other formal

characteristics - such as the use of prepositions in the case of English, of postpositions in the case of Japanese - correlate with the dichotomy of word order.

After all, it can't hurt to consider languages other than Indo-European before talking about General or Universal Grammar. In Chinese, for example, the position of a word within a sentence may determine whether it must be understood as a verb or a noun. In English (and other Indo-European languages as well) this difference is not caused by position but by designation (**to** club/ **the** club = mace). This elementary example shows that nouns and verbs must be understood as purely formal categories that belong to the *level of symbolic representation,* and that in each language different concrete contents of meaning are put in such formal slots - which means that there are no verbs or nouns as such. We must speak of the Chinese verb, the Chinese noun; the English verb, the English noun, etc.). Only actions, substances, qualities etc. are universal categories.

Not having recognized this basic distinction must be seen as the fundamental error of Chomsky's theory, which Pinker repeats when he equates the one with the other: "babies are designed to expect a language to contain words for kinds of objects and kinds of actions - nouns and verbs." (153). No, it's much more complex than that: "assault" is a noun on the formal level but it is an action and not an object on the pre-symbolic level. It is a characteristic of some languages that actions may formally be classified as nouns. I try to show why some languages resort to such "rank lifting" when they classify actions together with substances as nouns (the tree, the house... ; the assault, the leap...).

Meaning, Form and Formal Realization

The General Structure of Meaning
By this name I understand the logical and the informational parts of meaning in so far as they are common to all natural languages. Such a common basic structure would not exist unless the human mind dissects reality in basically the same way everywhere. There is no language where Action and Quality Syntheses do not exist or Statement versus Question and Command (Request) or the distinction between information known versus unknown. Why the human mind when faced with reality thus performs basically identical operations is a problem concerning psychology. Linguistics simply accepts the fact while at the same time insisting that apart from such basic operations the field of meaning allows for almost infinite variations. The distinction between Statements and Questions, for instance, may by blurred (you possibly know that...), the borderline between Substances and Actions may be indistinct in some cases (is the flash of lightning to be classed among actions or rather among static phenomena like house, tree etc.?). Reality presenting so many border cases, semantic classification may take different decisions in a non-finite number of specific cases. But this is no argument against the existence of a General Structure of Meaning as illustrated by the above-mentioned types of synthesis and the basic requirements of information.

Meaning: Its Logical and Informational Parts
There is no language which does not distinguish Substances from the range of possible Qualities like:

The tree is big/ small
The man is strong/ weak or:
The sea is deep/ shallow

Nor does any language exist which does not put this logical scheme to the requirements of information.

Meaning, Form and Formal (Symbolic) Realization

Somebody may want to say 'The <u>tree</u> is high' as an answer to the question, 'What is the object that you designate as high'. So, the Novum is represented by the term 'tree'. But when answering the question 'How do I recognize that tree? The information conveyed would instead be 'high'. The tree is <u>high</u> (not low). Now 'high' assumes the role of Novum.

The Logical Structure of Meaning provides the universal foundation of the analysis of reality by the human mind. In fact, what I call Syntheses like the Action Synthesis (Peter runs) or the Quality Synthesis (The tree is high) *reverses the previous mental analysis by reassembling its semantic units*, that is semants like Peter, tree, house, run, high etc.

The Resulting Logical Structure of Meaning consists of a rather limited number of Syntheses or admissible semantic patterns, which being universal permit all languages to be translated into each other (but never perfectly: below the level of universal patterns they exhibit an infinite variety of semantic concreteness).

Semantic universality and infinite semantic concreteness do not only characterize the Logical Structure of Meaning, the same opposition persists in its Informational Counterpart. Here too we easily hit on a number of universals. The distinction of Statement, Question and Command (Request) is a basic requirement of all languages and so is that of Novum and Topic.

Side Note: The Evolution of Language

But only developed languages make use of the Bound as distinguished from the Free Synthesis. Rank Lifting and Semantic Effacement too are only found in highly developed languages. No language ever fell from the sky fully developed but was gradually evolved over thousands of years. And written language may, of course, be much more sophisticated than its spoken counterpart (just read linguistic journals!).

This fact is denied by Chomsky when he postulated a fictitious "language module", with which all humans, regardless of culture, are allegedly equipped, so that there can be no question of an evolution of language as otherwise characteristic of all human organs and abilities. "For Chomsky, then, the basic justifications for saying that the capacity for language must be an innate module or organ, a computational mechanism, was the argument from the poverty of the input together with lack of correction, and ease of acquisition in childhood" (Pinker).

Meaning, Form and Formal (Symbolic) Realization

Unfortunately for the truth of his statement, Chomsky did not take the trouble to verify it by means of the abundantly available empirically material. In an article titled "So all languages aren't equally complex after all" (2018),[5] Christopher Hallpike quotes some of this evidence to prove that much of what can easily be formulated in developed languages is not expressed in the language of many tribal people such as for instance the Piraha: /They have/ "no relative pronouns; only single modifiers; only one possessor; no co-ordinates such as 'John and Bill came today'; no disjunctions e.g. 'either Bob or Bill will come'; only one verb and one adjective in a sentence; no comparatives or superlatives; no counting; no distinction between singular and plural; no quantifiers – some, all, every, none; nouns have no prefixes or suffixes; no color terms; no passive constructions; word order is not strict; no phatic communication (no greetings or farewells, 'please' or 'thank you' etc.)."

This diversity in cultural development reflects the ontogenetic process where children start with an incipient stage of elementary linguistic competence. "For instance, 'doggie' can be used to mean 'there's the doggie', 'where's the doggie', 'that looks like a doggie', 'I want the doggie', 'doggie pay attention to me', and so on...' as Jackendorf and Wittenberg have pointed out.[6]

The present work deals with Universals in Meaning, Form and Formal Realization without, of course, denying development. It merely asserts that development basically takes the same direction in all natural languages - which explains why we can translate and understand them. Form is biologically determined by the physiological properties of the human larynx and ear while the analysis of reality by the human mind is neurologically determined by the human brain. Neurologists assert that the latter acquired its present shape at least as long as 50.000 years ago. This explains why children as well as cultures follow definite lines of linguistic development from simple to complex. The present outline of Universal Grammar does not retrace this path but dwells on the basic pattern of that - presumably final - stage language has reached at our time.

---

[5] https://www.academia.edu/36569346/So_all_languages_arent_equally_complex_after_all

[6] Jackendoff, R., & Wittenberg, E. 2014. 'What you can say without syntax', in Measuring Grammatical Complexity, eds. Newmeyer & Preston, 65- 82. Oxford University Press.

Meaning, Form and Formal (Symbolic) Realization

Form and Formal Realization

The physical human apparatus producing sounds (the human larynx, tongue and mouth) as well as the organ receiving them (the human ear) define the range of possible sounds at our disposal. That is why the binary system of computer language with only two signs (+ / -) cannot be used as a means for creating a natural language. What I call the "Differentiation Value" (the common measure applicable to both Meaning and Form) just does not allow this to happen. In other words, the transformation of purely mental units like semants into measurable physical sounds (or their graphic representation on a sheet of paper) proceeds in quite a definite way in natural languages. It is arbitrary only in so far as any definite semant may be formally realized by any randomly chosen formant (as was already stated by de Saussure with regard to signifié - semants - and signifiant - formants), but beyond this level there are constraints or even Laws that delimit the range of the arbitrary.

Laws and Variety

The aim or purpose a General Comparative Grammar wants to achieve is to open up two perspectives at the same time. On the one hand, it wants to delimit the range of mere chance by proving that there exists a definite number of Universals on all three levels: The Logical and Informational Structure of Meaning and finally in the constraints governing their Realization in Form.

On the other hand, we are faced with contingency leading to nearly infinite variety. Constraints and laws describe the limits of freedom while variety sheds its light on the vast realm where the human mind operates and creates without any restrictions.

Some cautionary remarks

Difficulties arising from Meaning

Let me add some further remarks concerning meaning. Throughout the more recent history of linguistics, the obstacles to an adequate dealing with meaning have come from two different sides that by interest and temperament have been mostly opposed to each other. Let me call these the *logicians* and the *poets* of linguistics. The first were inclined to look for what is common in language, and they are easily led to discount its far-reaching differences. Here we meet the Universalists and those who are in love with formalism for the sake of formulas. The logicians in the study of language descent from Aristotle and the Scholastics, they were prominent among the theorists of Port Royal and they have been important in the most influential type of North American linguistics.

The *poets* among the students of language were primarily found in the romanticist school of German linguistics. Herder initiated a movement, which was subsequently carried on by Trendelenburg and Wilhelm v. Humboldt. In recent times Whorf has given new life to this movement in America, while in Germany this current of thinking has always had its outspoken followers (Weisgerber, Weinrich and others). The contributions of both these currents can hardly be overestimated. *Logically* minded linguists insisted on method and exact analysis, they rejected intuition if not substantiated by empirical demonstration. Their chief aim was to apply to and to make available for the study of language the predominant standards of science. In recent times this tendency has its most prominent proponents among North American linguists from Bloomfield to Harris. Distributionalism as an exact method for describing the outward form of languages was the natural outcome of their works.

The romanticists or poets among linguists were sensitive to quite other dimensions. They clearly perceived what the logicians had left out of their schemes, namely the entire range of phenomena that are unique in each language, that is, its tremendous wealth of peculiar semantic concepts and unending formal variety. They easily discounted the early attempts at universalism, as proclaimed by Port Royal and its followers, by showing that the familiar concepts of linguistic description were fundamentally biased for they had exclusively been taken from those few languages the linguistic observers happened to be familiar with. It is to this current of linguistic thinking that the science of language owes the enthusiasm for diversity as expressed in foreign cultures and foreign languages.

And yet both currents tended to be one-sided. By putting exclusive stress on their particular point of view, they were prone to ignore the other half of the truth. And here we come back to the problem of meaning. The logicians belonging to the North American school found meaning far too intractable for exact analysis; they therefore concluded that it had no place in the science of language from which it should be completely removed. According to their point of view, only linguistic form represents physical objects (that is, organized acoustic sounds or their representation on paper) that we can measure. Form may be described in quite unequivocal terms, so as to be susceptible to clear definition.

Meaning is neither physical nor can it be measured. Dealing with meaning therefore seemed to be a hopeless and elusive undertaking. It has no natural boundaries and its shades and interlingual variety verge on the infinite. Viewed in this perspective the opposition to making meaning an object of science seems understandable. We know that the infinite has no place in science. *Science cannot come to terms with the boundless or indefinable. That is why the logicians decided that meaning had to be discarded* as, indeed, was to happen in Distributionalism.

The poets of the linguistic tradition were in no danger to fall into a similar trap. For meaning much more than mere physical form was their chief object of interest. As a matter of fact, they succeeded in sharpening our eyes for its complexity. But - paradoxically - the result of their endeavors turned out to be the same as that of their opponents. To all practical purposes *they managed to push meaning out of science*. Nor is it difficult to understand why this happened. They tended to look at meaning exclusively in the particular shape and individual connotations it assumes in each language. A sentence like English 'He will give it to you' had to be viewed as something entirely different according to whether it was expressed in Navaho, Paiute or Japanese[7]. So, the very concept of Meaning as something meaningful beyond each single language was all too easily dismissed as a mere construct. To be sure, *this approach was comparative from its very start but it achieved the paradoxical result of making comparison useless*. If indeed we meet with the individual and nothing else in each language, then comparison is without foundation – it lacks any tertium comparationis.

Logically speaking, the position of those, whom I call the poets of linguistics, is, of course, self-defeating, as it is contradicted by the very fact of translation. It leaves the question unanswered what it is that makes us speak *of the same or at least very similar semantic contents when making or comparing translations*.

Meaning - the Core of Language

It is time to overcome the difficulties confronted by 'logicians' and 'poets'. Meaning is the core of human language, to convey meaning is the very reason why languages exist in the first place; any theory that abstracts from meaning does, at best, deal with a skeleton. Such an observation seems true by intuition. But intuition is not enough. It is with regard to method that valid arguments have to be found.

Let me, first, show that the rejection of meaning because of its boundlessness is not at all justified. When we state that an element of meaning, such as for instance English "tree", is formally realized in quite different ways by different languages (e.g., 'ki' in Japanese, 'tree' in English, 'Baum' in German, 'mu' in Chinese, 'arbre' in French, etc.), we do not want to say that this concept is semantically identical in all these languages. On the contrary, within each language under consideration we know it to be quite different in semantic *shades and associations*. Our comparison (and such comparison is

---

[7] An English sentence like 'he will give it to you' appears in Navaho as 'thee-to-transitive-will-ROUND THING IN FUTURE', in Paiute as 'GIVE-will-visible thing-visible creature-thee (cf. Sapir in Hymes (ed.)1964:103). In Japanese we would have three alternatives 'give to you' (equal to equal)', give to you (lower to higher), give to you (higher to lower).

I have taken the first of these examples from Sapir though Sapir is certainly not a typical representative of what I call the 'poetic current' in linguistics. On the contrary, Sapir is a comprehensive thinker of great persuasiveness, who reunites what is best in both traditions.

at the base of every act of translation) does not rely on the full concept but on its *distinctive traits*. We limit comparison to that minimum of characteristics that make the concept "tree" stand out as something distinct from other concepts like "stone" or "cloud". This is the way comparative linguistics will have to precede when defining that General Structure of Meaning, which remains the same in widely different languages like Japanese, English, French or Navaho. Translation is logically possible only because it is based on such a Tertium Comparationis.

To take a very adequate term used by Chomsky, to which, however, I will give a different content: There must needs be a Semantic Deep Structure hidden behind those differences of semantic conceptualizations. The Logical and Informational Structures of Meaning, *composed of Semantic Categories, Types of Synthesis and their Informational Use*, is meant to give flesh and bones to this basic dimension of language.

Difference between linguistics and logic

Logic interprets the copula 'is' as expressing identity. Linguists, however, must conclude that there is no constant correlation between the copula and any function determined by logic. In the two sentences:

a) London is the capital of England   or
b) London, the capital of England, (is a good place to live)

the relationship between 'London' and 'the capital of England' is one of identity though the copula is not be found in (b). For this reason, the copula 'is' cannot primarily serve to realize identity. It must have quite different functions too, namely purely linguistic ones (see IV,6).

While the copula 'is' may be absent at places, where identity is expressed, it may be found, where no such relation obtains, as in 'The tree is green'. Obviously, the Substance "tree" and the Quality "green" cannot be identical since the latter only represents a modification of the former. But again, we are confronted with the opposition:

a) The tree *is* green
b) The green tree (is my favorite).

Some Cautionary Remarks

The real opposition which is not a logical but a purely linguistic one will be explained later. In this case, it concerns the difference between the Free and the Bound Synthesis thus belonging to the Informational Structure of Meaning (see III,4).

Meaning and its Realization in Form

The present study of meaning is meant to fulfill a specifically linguistic task. It will therefore be different from similar investigations in philosophy or in psychology. The object of comparative linguistics is to show how meaning is associated with form in some given language or, as I will call it, how it is 'realized'. We just spoke of "tree" as a concept of pure meaning, different from its formal appearances ('ki', 'arbre', 'mu', 'albero', etc.). It is, indeed, of crucial importance that meaning in comparative linguistics be always understood as *pure meaning*, which means that no trace or remnants of formal characteristics are allowed to creep into its definition. This is no less than a logical prerequisite, as the main question of comparative linguistics concerns the *distinct ways different languages associate meaning with form*. If we were to allow any reference to form in our definition of meaning, our answer would become circular. We would indeed ask how meaning determined according to formal criteria appears (is realized) in form.

Logically it seems to be evident that the basis of comparative linguistics should be pure meaning. However, it is one thing to make a statement of principle, while it is quite another matter to embark on its practical implementation. Let us be a little more factual and show what we have in mind when speaking about meaning defined with reference to form. The difficulties we necessarily hit upon will then immediately come to the surface. Terms like noun, verb, adjective etc. are defined both with reference to form *and* to meaning. They designate specific formal classes in some specific language. In other words, *most of the central concepts* hitherto used in the science of language are ambiguously defined.

This is a legitimate procedure in Traditional Pedagogic Grammar as it renders description so much easier. But it can have no place in Comparative Linguistics. For this reason, we are compelled to reject such terms as descriptive units within pure semantic structure.

Some Cautionary Remarks

Old difficulties removed

Once the analysis of language is based on a general structure of meaning, some typical difficulties arising in traditional linguistics do no longer occur. Let me illustrate this by means of some randomly chosen examples. Consider the two sentences 'He gets up when the sun rises' and 'He gets up at sunrise'. Traditional as well as hybrid (Chomskyan) grammar, basing their descriptions on form, have treated these sentences as if they were different entities. The two expressions 'when the sun rises' and 'at sunrise' appear under different headings and were given different names. This procedure not only prevented traditional grammar from accounting for the common semantic ground within one and the same language, it furthermore stood in the way of comparing different languages. Comparative linguistics cannot abstract from common semantic structure, it must describe the latter *independently* from any formal appearance. The common semantic structure in question (a temporal specification) is *formally realized* in one example as 'when the sun arose' and in the other as 'at sunrise'. As a matter of theoretical insight this conclusion seems to be fairly obvious, but things turn out to be rather difficult as soon as we try to define what we intuitively mean by "common semantic structure". Traditional and hybrid grammar have hardly developed any terms of a purely semantic nature - it is always the particular *form* on which, in definition and coining of names, they used to base their terminology.

Take another example, the so-called "relative clause". In English we may say 'the man, who goes home, is my friend' or 'the man going home is my friend'. In Japanese and Chinese, only the second type of formal pattern may be adopted. Obviously, both alternatives represent the same basic semantic structure but they are treated in traditional grammar as distinct and are therefore described by different terms. Again, traditional and hybrid grammar take form as their measuring rod. This leads to typical difficulties. If 'who goes home' is a relative clause (because characterized by a relative pronoun) and '... going home' is not (because it lacks a relative pronoun) then, obviously, Japanese and Chinese have no relative clauses though they have no difficulty at all in formally realizing the underlying structure of meaning.

Let me conclude: No theory of linguistics can be said to be complete if it does not explain the principles common to these and other formal alternatives. In Universal Grammar, as here proposed, the fore-mentioned alternatives are treated as so many different formal realizations of identical semantic structures. Difficulties disappear as soon as we start from the General Structure of Meaning and then ask for *alternative ways* of formal realization.

Once clearly stated, this again seems to be a fairly obvious theoretical conclusion but traditional grammar provides no help to guide our procedure. For obvious reasons, we cannot give the name of "relative clause" to the underlying semantic structure - such designation only refers to a particular formal pattern (using relative pronouns). We are obliged not only to define exactly what we mean by "underlying semantic structure" but also to create entirely new names. Ready-made names simply do not exist in present-day linguistics. Thus, I had to define the common semantic structure according to purely semantic criteria and give an adequate name to it ("bound" or "non-information" synthesis). This has been one of the major tasks to be completed.

Indeed, in developing a purely semantic structure I had to start from scratch. This may explain why, in my first dealings with this subject, I rigorously discarded traditional terms because any term defined according to formal characteristics leads to tautology or what I call "hybrid grammar". But I had, so to speak, thrown out the child with the bath. The old terms may still be quite useful when applied where they rightfully belong. So, it is legitimate to say that the "bound action synthesis" is formally realized in English by means of two formal patterns one of which we may designate by the familiar name of "relative clause". Using traditional terms for the purpose of naming *specific formal* patterns (and only these) saves us the trouble of looking for neologisms. As a rule, new terms should only be used if they are imperatively demanded by new requirements of theory.

Explanatory and Descriptive Grammars

Descriptions of language have been called grammars but these may serve quite different tasks. *Pedagogic*, *normative* and *comparative* (universal) grammars may be said to represent the mainstream of linguistic description. However, when dealing with language, description can only be one among different objects of investigation. Explaining the why and how of languages is a further and more ambitious task.

Pedagogic grammar tries to account for the salient features of some given language. In doing so it conforms to the practical imperative of conveying a maximum of information on a minimum of space. A pedagogic grammar of Latin and a pedagogic grammar of Chinese will therefore be quite distinct. Their descriptive terms must be adequate to the language described, not to any scientific imperatives valid for the comparative study of language. However, when people first tried to describe their own language, for instance French or German, they used the terms derived from the grammar of Latin in order to understand the unfamiliar by means of familiar terms. This procedure is legitimate because induced by the specific purpose of pedagogic grammar, but it becomes an obstacle to the progress of linguistics as soon as language is viewed as the object of *scientific* investigation. The terms of description used in pedagogic grammar turn out to be largely useless in a comparative treatment of language.

Normative grammars represent the first outcome of man's study of language. The famous grammar of Panini was meant to provide once and for all a compendium of rules for the correct use of Sanskrit. When language was thought to represent an instrument of divine origin, it had to be protected against the onslaught of time and the influences of corruption. Grammar was understood to direct and impose the right use of language, it was not conceived - as we understand it today – namely, as a faithful and economical description of language or as a means of explaining it. Normative grammar devised its basic terms in view of descriptive economy, not in view of scientific investigation.

The Explanation of Language centers around two basic questions. The first is concerned with single languages. What makes a speaker capable of *producing an infinite number of sentences*, which are considered correct by the native linguistic community? There must be rules at his disposal allowing him to produce a potentially infinite number of right utterances and to discard the remaining infinity of wrong ones. The *speaker's generativeness* is the first dimension of language to be explained. This question would not have been asked without the seminal ideas of Noam Chomsky.

The second question concerns language as such. What makes the human mind capable of *producing an infinite number of languages*? There must be rules allowing it to discard all those, which could be realized as artificial languages but (for various reasons) not as natural ones. This time it is the *general linguistic generativeness* of the human mind that has to be explained with regard to its origins and working. This question cannot be asked let along treated in an adequate manner without rejecting much of what Chomsky has offered as his own solution to the problem of generativeness.

In contrast to pedagogic grammar, it cannot be the purpose of a *comparative study* of language to *describe any particular language in the intuitively most easily understandable way*. Comparative grammar aims at describing and explaining the structure and composition of language in general terms applicable to all natural languages. This more extended perspective results in a much greater strictness of method. It is impossible in General Grammar to choose definitions according to expediency. We have shown that comparative grammar must resort to a set of purely semantic terms (as embodied in the logical and informational structure of meaning) adding to these a set of purely formal ones.

Some Cautionary Remarks

Analytic and Constructive Linguistics

Our previous analysis explains the genesis of language in the following scheme consisting of three basic parts:

1) Meaning (semants, syntheses etc.) is formally realized by means of:

2) definite formal units = free or bound formants (independent words or dependent affixes) together with position, intonation and tones in:

3) structured formal sequences, the most elementary units of which we name 'sentences', as these normally represent what in the sphere of meaning are the basic 'Types of Synthesis' modified according to the needs of information.

Now, the linguist may analyze any specific language with the aim of describing how the General Structure of Meaning (Pinker's 'Mentalese') is embodied by means of given rules of formal realization in a particular language. Or he may ask the more general question what formal solutions are theoretically possible to create different natural languages - *a problem unconsciously solved by any newly arising linguistic community*.

Both approaches are quite different. In the first case, a child learns to embody Mentalese in formal rules it takes over from an already existing linguistic environment. In the second case, we see how mankind - using the set of formal means at its disposal - makes certain formal choices in order to create language - real and possible ones - but without being guided by existing patterns of formal realization. Thus, Comparative Linguistics necessarily consists of two different compartments, namely "Analytic Linguistics" and its counterpart "Constructive Linguistics".

Whenever a new language was created by men in the course of history they had to make certain unconscious decisions. They could, for instance, choose four different syllables like ben, ban, bun, and bin in order to realize four different concepts (most languages follow this pattern) or choose just one syllable, for instance ben, and modify it by means of four different tones (ben^1, ben^2, ben^3, ben^4) as does Chinese which for this reason cuts the number of needed syllables down to at least a quarter.

Another choice had to be made according to whether substances and actions should be distinguished by designation as in most Indo-Germanic languages (Latin: puell-*a*, puell-*ae* /ama-*t*, ama-*nt*) or by mere position as in Chinese. The resulting formal slot of substances in Latin called nomen$_{Lat}$ is exclusively defined by designation while its Chinese counterpart (nomen$_{Chin}$) is so exclusively by position. And of course, a similar choice had to be made with reference to agent and patient. In Chinese and to a substantial degree in English too these are distinguished by position while in most Indo-Germanic languages (puell-a/ puell-am) as well as, for instance in Japanese, designation serves the same purpose ("wo" for patients). The genesis of language consists in such choices between admissible alternatives.

1 Explaining Generativeness

The first of these compartments deals with natural languages still in use or found in historical documents. It explains the *particular generativeness* of a speaker of English, Chinese or any other definite language. The second describes *general generativeness*, namely how natural languages are developed by human beings when - obeying to the constraints of formal realization - they create *totally new idioms*.

Constructive Linguistics endeavors, first, to describe the field of arbitrariness in natural languages, while, in a further step, it aims at specifying the *limits* of arbitrary variety, that is, the constraints presented by formal realization, which may lead to *definite laws*.

Taken together, Analytic and Constructive Linguistics cover the entire field of Linguistics. But in order to be complete, the above scheme has still to be widened so as to include Paratax.

In all natural languages, the formal arrangements of basic units (formants or words) occurs on two different levels. The synchronic level describes the *succession of formants* within a syntactic structure (Syntax), the paratactic level (Paratax) describes the *class of formants* allowed at each position in the syntactic chain, for instance English nouns, English verbs, English adverbs etc. In this way, the genesis of language presupposes a syntactic and at the same time a paratactic ordering of elements.

Analytic and Constructive Linguistics

The above scheme must therefore be widened so as to include the latter:

1) Meaning (semants, syntheses etc.) is formally realized by means of:

2) definite formal units = free or bound formants (independent words or dependent affixes) together with position, intonation and tones) in:

3) syntactically structured formal sequences representing the basic 'Types of Synthesis' modified according to the needs of information. These formal sequences are called 'sentences' consisting of subunits (formants) that paratactically organize semantic classes into formal ones (traditionally called English, Chinese, Japanese etc. nouns, verbs etc.) in a language-specific way. Their syntactic order is again language-specific (so that English syntax is quite different from its Chinese counterpart even when both realize identical structures of meaning).

This procedure defines the structure of the present work. It starts with chapters II and III expounding Meaning and its appearance in natural languages. They comprise the Types of Synthesis and the use made of these for the requirements of Information. Chapter IV then describes the formal means at the disposal of natural languages and how these may be used to translate purely mental images (Meaning) into acoustic waves or the corresponding graphic symbols. Chapter V stresses the fact that formal ordering is not limited to Syntax - Paratax is as important a factor when describing linguistic Variety. The last chapter is a modest attempt to delimit the range of contingency showing that language - every natural language - is subject to certain laws. Hopefully, this field of research will be widened in the future.

2 Concerning method

This paragraph is a mere repetition aimed at further emphasizing the difference between traditional terms and those to be used in Universal Grammar. Pedagogic grammar will strive for the most succinct description. It may therefore use terms only applicable to the language it is meant to describe. When defining the English noun, we may do so exclusively by referring to its formal surrounding without any reference to meaning. Most English nouns may thus be defined as being modifiable by the suffix '-s' indicating plurality or by their appearance after the determinants 'a' or 'the'. As for

English verbs, a majority of these must allow for tense determinants like '-ed' for past etc.

Obviously, such *form-specific* description cannot serve for comparison. Comparative Analytic Grammar must therefore rely on purely semantic definitions. We find that nouns in different languages have one semantic feature in common: Invariably, they comprise (animate or inanimate) Substances like man, house, wall etc. Likewise verbs in all languages have one common semantic feature: they comprise Actions like "walk", "hunt", "climb" etc. But due to different *paratactic* classification English, Chinese and Japanese nouns comprise different semantic elements, and so do verbs, adverbs etc. Any comparative description that abstracts from such differences does not explain linguistic variety but, on the contrary, explains it away.

3 Concerning the use of traditional terms

However, we don't contradict the requirements of sound method as long as we use our terms for a particular purpose that does not deny existing variety. We may for instance ask why a language like Japanese places *the verb* at the end of phrases and uses postpositions, while English uses prepositions and places *the verb* between agent and patient. When we restrict our analysis to a particular problem we may abstract from the fact that English and Japanese verbs have by no means identical semantic contents.

Likewise, we may use terms like "relative clause" and "pronoun" though these are not general terms. I already mentioned that there are languages like Chinese or Japanese that do not have "relative clauses" headed by a "pronoun". These terms remain useful nevertheless if we want to designate a certain pattern of formal realization.

4 Formal relevance

The range of meaning (or Mentalese) is potentially infinite as is reality itself. But the human mind dissects reality in specific ways that are finite (the *Types of Synthesis*). Even so, the analysis of Meaning may be carried on to deeper and deeper levels demonstrating at the same time the common procedure of the human mind in whatever language it is expressed, and, on the other hand, unending variety. Unending description based on unending variety cannot, however, be an end of science - or else scientific description would be coterminous with the object described. We therefore ask for that parts of Meaning (in its Logical and Informational Parts) that have *formal relevance* as they either impose definite formal means or only allow for a limited number of formal alternatives. Formal relevance leads to the formulation of laws while its absence describes the very opposite: the range of the incidental.

5 Synthetic, Agglutinative, Fusional, Polysynthetic Languages

In so far as these four alternatives to the isolating type (for instance Mandarin) merely concern formal realization, where closed-field semants are realized by means *of bound formants* (see VI,5), they represent an old fashioned classification of language that does not tell anything about their underlying semantic cosmos.

But bound formants expressing closed-field semants may indeed give rise to entirely different views of reality. They may either be frozen and meaningless - as gender in modern German with feminine forks and masculine spoons - or represent a philosophy of reality - as gender did at the time when it originated in Indo-Germanic languages. In other words, even identical formal patterns may -in the mind of speakers - be filled with entirely different semantic content and overall importance. Sapir's and Whorf's far-reaching interpretations may be relevant in some but definitely not in all cases.

Analytic and Constructive Linguistics

I Basic Terms of Universal Grammar

We need a new vocabulary which defines basic terms interlingually so that they have the same content regardless of the language to which we apply them.

A) Basic Terms referring to Meaning:

1. Logical Structure of Meaning:
Semants = smallest semantic units. Open field/ closed field semants
Categories of Semants = Substances, Actions, Qualities etc.
Syntheses = logically complete associations of semants
(so named because the synthesis reverses the previous process of how the mind *analyzes* reality)
Types of Synthesis
Agent/ Patient, Possessor etc.
Enlarged Synthesis
Open field/ closed field syntheses
Connections = free + bound Syntheses
Conjunctions = combinations of free Syntheses
Ranks
Semantic Inversion
Frozen Syntheses

2. Informational Structure of Meaning
Command (Request)
Topic/ Novum
Statement/ Question
Free/ Bound Syntheses
Semantic Explicitness/ Semantic Effacement/ Total Semantic Effacement
Rank Lifting
Informational Shifting

B) Basic Terms referring to Form:

Formant (mono- or polysyllabic) = smallest formal unit for expressing a semant
Tones = creating new formants by tonal modification
Intonation = modifying a group of formants
Position = order of formants
Sentence = smallest formal unit realizing a synthesis, a connection or a conjunction

C) Basic Terms referring to both Meaning and Form

Differentiation Value (Dif-Val) =
the Tertium Comparationis allowing Form (acoustic sounds, visual signs, gestures, electrical charges) to become a means for the expression of Meaning.
Differential Analysis (79, 119)
demonstrates how identical elements of meaning (semants) are expressed by differing elements of form (formants). In its intuitive or methodological application, differential analysis represents the only instrument of intra- and interlingual comparison, since it replaces the arbitrary signs of any given language with the meanings they are designed to express.

Most terms of Traditional as well as Generative Grammar are not universally applicable. This is true of concepts like Verb, Noun, Adjective etc. as these are characterized by their specific *paratactic content* in any given language. In other words, the class of English verbs (V_{engl}) differs in semantic content from the class of Japanese (V_{jap}) or Chinese verbs (V_{chin}) though these contents do, of course, overlap - the reason for using the common term "verb" in the first place. Relative pronouns or relative clauses as well as the so-called passive voice are even less universal - they are unknown in certain languages. Seemingly simple and intuitively understood concepts like subject and predicate are ambiguously defined while Chomskyan terms like recursion and embedding do not further our understanding of language (49). All this will be shown at the proper place.

D) Basic Terms referring to Formal Realization

Formal Relevance
Formal Syntax/ Formal Paratax
Formal Equivalence/ Abundance (Redundancy)/ Deficiency
Formal Ambiguity
Formal Extension
Free/ Bound Formants
Formal Ellipsis (66, 71, 96)
Zero-Form (60, 71, 84, 101, 122)
Formally induced Semantic Tingeing (103)

Basic Terms of Universal Grammar

II The Logical Structure of Meaning

1 Semants – the Subunits of Meaning

If the preceding assumptions are correct, the first task of comparative linguistics must consist in a definition of Pure Meaning, namely its units and subunits.
Subunits consist of basic semantic elements like tree, stone, car, wolf, man etc., which I will call **semants**. Taken separately these do not, however, convey any information. Obviously, lexical semantic items may not be considered units as they fall short of the *requirements of information*. Nobody converses in the following way:

Cold, Peter, volcanoes, serenity, green etc.

Semants represent nothing more than semantic items abstracted by the human mind from surrounding reality. In order to be of any use in information they must combine in a *specific way* with other semants. The result is not what traditional grammar (including its Chomskyan variant) calls a sentence - sentences are meaning realized on the level of sounds (or their graphic representation). Since we made it our task to adhere strictly to the semantic level, a different notion must be found. I will call these basic units of information ‚Syntheses'.
In all human languages, we only meet with quite a restricted number of syntheses. The most elementary one being the Action and the Quality Synthesis, where a substance is modified either in a permanent or in a temporary way.

The Logical Structure of Meaning

2 Syntheses - the Units of Meaning

Structure of meaning	*English realization in form*
Quality Synthesis: Tree, green	The tree is green Or: The green tree...
Action Synthesis: Man, run	The man runs (is running) or: The running man...
Identity Synthesis: Socrates, man	Socrates is a man or: The man Socrates...
Part-of-Whole Synthesis: house, roof	The house has a roof or: The roof of the house...

The dichotomy of the synthesis "The tree is green" versus "The green tree..." etc. will be dealt with in chapter III: **The Informational Structure of Meaning**.

The modification of substances by qualities or actions **allows for degrees**:

Structure of meaning	*Possible realization in form*
Quality Synthesis: Tree, green, hardly	The tree is hardly green
Action Synthesis: Peter, run, fast, very	Peter is running very fast

The basic synthesis may still be enlarged by either **spatial or temporal specification:**

Structure of meaning	*Possible realization in form*
Quality Synthesis: Tree, green, here, now	This tree is green just now
Action Synthesis: Peter, run, yesterday, London	Yesterday, Peter was running through London

The types of synthesis are not restricted to natural language but used in formal logic as well – the latter being nothing else than a subset of the former. Natural languages do, however, extensively use types of synthesis not to be found in logic. They do so, for instance, when referring outward events to the psychic state of a human observer: The **Psychic-state Synthesis**.

Structure of meaning	Possible realization in form
I, see X (X = Tree, green)	I see that the tree is green
I, believe X (X = Peter, run, fast)	I believe that Peter is running fast

The Logical Structure of Meaning

Furthermore, natural languages universally introduce certain other types of synthesis, which denote basic social facts, as does the **Possessive Synthesis**:

Structure of meaning   *Possible realization in form*

Peter(Pr), house(Psd), own   Peter owns a house
                                      Peter's house (is quite big)

When used in the linguistic context, possession does not, of course, imply any ideological point of view - it may simply indicate a special relation of responsibility of a human being (the possessor) for certain objects possessed. In many (possibly in most) languages, the possessive synthesis is formally realized by way of shortcuts like the English 's' following Peter. While possession has no role to play in formal logic it has thus a high *formal relevance* in the logic of language. In many, probably in most languages formal realization of the possession synthesis follows that of the part-of-whole synthesis: Though based on an entirely different semantic relation, Peter's house is in form identical with Peter's head.

3 Quantitatively Enlarged Synthesis - Ranks

The elementary Action or Quality Synthesis may be modified in a particular way characteristic of every developed language. Jespersen referred to this fact when he spoke of 'ranks'. The process of mentally analyzing reality results in a definite ranking order of semantic specifications.

I	II	III	IV
lion	run	fast	very
tree	become green	fast	very
tree	green		very

The Enlarged Synthesis permits of four ranks, the third rank being admissible only if the second is a process in time and not a static quality.

Note that the definition of ranks conforms to purely semantic criteria. Further semantic specifications such as "tree, green, very, not, presumably" ('the tree is presumably not very green') do not belong to the enlarged synthesis, as the semant "presumably" is part of an altogether different synthesis

The Logical Structure of Meaning

(subjective synthesis): "I, we, etc. presume, X (where X = tree, green, very, not)".

Semantic ranks are important as a means of understanding how the process of mentally *analyzing* reality is inverted when leading to the types of *synthesis*. It is the more particular, which becomes the subject of specification by the more general.

We find this scheme at the bottom of quite different types of synthesis such as the following:

quality synthesis:	tiger,	yellow	(the tiger is yellow)
identity synthesis:	Socrates,	man	(Socrates is a man)
action synthesis:	cat,	jump	(the cat jumps)
localizing synthesis:	object,	here	(the object is here)

Whether it be a tiger or a thing in general that we pronounce to be yellow, in each case this statement implies that a substance (a sum total of qualities and possible dynamic modifications) becomes characterized by just one or a few among these qualities or states. In this way, the more particular (the sum total) becomes specified by the more general (one instance or a few out of a sum total). "Socrates" is more particular than "man" and so is "cat" with regard to "jump" - one of its possible modifications. As it belongs to the characteristics of any object to exist *somewhere*, this, too, must be considered a more general state than being an object as such. Conceptual ranking of this kind is already to be found in the nuclear synthesis (consisting of just two semants like "tree, green") but it is carried over into the modified one. Take, for instance, one of the above mentioned examples:

I	II	III	IV
tree	become green	fast	very

Here "fast" is one possible instance of the process "becoming green" while "very" is in its turn one such instance with regard to fast.

I will call 'logical head' a rank, which occupies a higher position with regard to its followers, and these will be called 'Logical colon'. Thus "tree" is logical head to "become green, fast..."; "become green" is logical head to "fast, very", while "fast" occupies the same position with regard to "very". The traditional division of subject and predicate when defined in a purely semantic way is partly based on the ranking of particular versus general.

The Logical Structure of Meaning

The analysis of ranking provides the criterion for distinguishing the modified synthesis from enlarged ones (spatial, temporal enlargement etc., see below) and from conjunctions (the combinations of different types of synthesis, see below). Some languages like English do not make any difference in formal realization:

Rank III enlargement	The Golem approaches rapidly
Temporally enlarged Synthesis	The Golem approaches now
Conjunction	The Golem approaches certainly

The conjunction arises out of the combination of two different types of synthesis (I, we, etc. are certain that X will happen (X = the Golem approaches).

The Quality Synthesis provides the semantic basis for comparison:

I	II	III	IV	
Peter	bigger (Paul)	much	very not	(Peter is not very much bigger than Paul)
Peter	bigger (all)			(Peter is the biggest of all)

More instances of the Enlarged Synthesis

NUCLEUS			ENLARGEMENT	TYPE
Bill(Ag)	run(s)	|	fast,again,twice...	monovalent action synthesis
		|	temporal	
Bill(Ag)	hit(s)	|	fast,again,twice...	bivalent action synthesis
Peter(Pt)		|	temporal	
Bill(Ag)	strike(s)	|	on the head	trivalent action synthesis
John(Pt)		|	local specification	
stick(Ins)		|		
Bill	(is) in	|	right now	localizing synthesis
(the)house		|	temporal	

The Logical Structure of Meaning

Just as there are borderline categories (like rain, flash of lightning etc., which may in some languages be formally realized like substances, in others like actions), so semantic roles too may sometimes be formally realized in alternative ways.

a) John(Ag) smeared the wall(Pt) with paint(Ins)
b) John(Ag) smeared paint(Pt) on the wall

The initiator of action is unambiguously determined: John is the Agent. The Patient is defined as any substance modified by the action. There is no doubt that this applies to the wall: Wall is the patient. However, the paint is modified too, it is therefore not unreasonable to classify it as a Patient. The synthesis then is locally enlarged (on the wall). English realization shows that the instrument of action, in this case the paint, is sometimes allowed to be formally realized at the same place as true Patients - an instance of *formal extension*. Nevertheless, the majority of events can be clearly ranged as substances or actions or as patients and instruments, even if some cannot. It is sufficient to base our investigation on clear-cut cases if we want to compare formal realization in different languages.

Enlargements may in certain instances be formally classified in different ways. In English, 'again' is classed as an adverb in English. There is no English verb 'to again'* but there is an expression 'to resume', which may, in some cases, be used with identical meaning. In its identical acceptance the semants "again", "resume" will in English be realized as either an English adverb ($_e$Adv) or as an English verb ($_e$V).

Formal alternative of an enlarged synthesis

a) they(Ag)   studied   Spanish(Pt$_{psy}$)   again
   eN+eNom   eV   eN+eAcc   eAdv

b) they(Ag)   resumed   the study(Pt)   of Spanish(Pt$_{psy}$)
   eN+eNom   eV   eN+eAcc   eN+eGen

The two English terms 'to resume' and 'again' may differ in semantic nuance but in some of their occurrences they can be used in a semantically identical way though they imply totally different formal constructions. In example b) 'Study' is turned into a pseudo-patient (*Pt*) while 'again' assumes the formal appearance of an action. Traditional grammar all but ignores such deep-lying semantic identities.

Semantically effaced Enlargements

All syntheses are susceptible of spatial and temporal enlargements because all events are determined according to space and time. If we say, for instance, 'Peter went to the theatre', we know that the event happened in Berlin on the 14th of December. The statement refers to some unique event. But such specification may or may not be present in the formal structure. Answering a question, we may simply say 'He went'. The specific spatio-temporal determination is, as a matter of course, present in the mind of both speaker and listener (because otherwise the event would be placed outside of time and space) but it is not spoken about (and therefore both *semantically and formally effaced*) because both know it anyhow. But temporal and/or spatial specifications must, of course, be realized in form if the listener does not dispose of this information:

"He, walk, home, yesterday"   'He walked home <u>yesterday</u>'

may be an answer to the question: <u>When</u> did he walk home? *Semantic effacement thus belongs to the Informational Structure of Meaning as it depends on the requirements of information.*

Psychic-state Syntheses

The mental analysis of reality into meaning leads to a basic distinction in types of synthesis. They may refer to the inner processes of a perceiving or apperceiving subject, or to outward events. I will call these two types 'subjective' and 'objective', respectively.

The types of synthesis listed above are all objective in so far as they do not involve human perception, apperception or socially established relations.

The Logical Structure of Meaning

Other types of synthesis do just that - so we may call them by the name of Subjective Syntheses:

Types of subjective synthesis

We cloud(s)	see	|	in the sky spatial	**perception** synthesis
We it	believe	|	not	**apperception** synthesis
He Betty	likes	|	always	**affection** synthesis
Peter house	own(s)	|	now temporal	**possession** synthesis

The subjective synthesis essentially differs from the objective one. The latter always represents an **open field**. Let us take for example the quality synthesis (S, q). We may substitute an almost infinite number of words both for S (house, tree, water...) and for q (high, flat, thick, hard...). But in subjective syntheses, both are possible: open and **closed fields**. Just as most Indo-European languages adopt closed semantic fields in order to express oppositions like singular/plural, masculine/feminine/neuter (*an essential trait in formal realization*), so certain Native American languages contrast the direct experience of a speaker with mere hearsay. Such differences can then be expressed by mere affixes.

    The preceding semantic classification may easily be carried on to deeper and deeper levels. There have been numerous languages that classified substances not merely according to gender but in view of their being animated or not, humans or animals, flying or swimming and so on. Similar distinctions can be made for actions. Possession is no action at all but a socially established relation between a human being and certain objects. And it is by no means self-evident that most modern languages use the same type of formal realization for possession and biological relations as in 'Peter's house' and 'Peter's mother' or 'his wife'. In other words, the above-mentioned classification could be endlessly pursued. Though the number of basic syntheses is quite small, *it certainly allows for endless subclassification*. The linguist as a poet is interested in just this infinity of possible semantic differentiation, especially in enlargements of the nuclear syntheses made obligatory as the expression of tense, number and gender in most Indo-Germanic languages. In other languages, for instance in Nootka, aspect plays a dominant role. Actions must be qualified according to whether they are durative, inceptive, momentaneous, graduative (similar to English progressive), pregraduative, iterative or iterative inceptive. In Japanese, the speaker has to express whether he consider the status of the person addressed equal, higher or inferior to his own. I will deal with obligatory types of expression in more detail later (see IV,5).

The comparative linguist points to these differences if he wants to elucidate the variety of mental analysis applied to reality. Anthropologists may even go further in explaining how the interaction with natural environment, social conditions and, more generally, a given stage of development favor certain ways of analyzing reality.

If, on the contrary, the linguist wants to elucidate formal realization, he behaves not as a poet but as a logician. In this case he is interested in clear-cut cases only, namely those to be met with in all languages - at least in all developed ones.

4 Semantic inversion

Semantic inversion describes alternative conceptualizations of one and the same event as exemplified by the two following statements ‚Mary is taller than Ann' or ‚Ann is smaller than Mary', where the same relation is alternatively defined with reference to the first or the second substance (persons in the present example). Such alternatives belong to the *Logical Structure of Meaning*. Semantic inversion is, however, easily confused with *Informational Shifting* of Topic and Novum (see below, III,3), which belongs to the *Informational Structure of Meaning*.

Any synthesis containing more than one substance is susceptible of *semantic inversion*. The relation existing between the substances can be defined with reference either to the one or the other. Consider the following example of a bivalent action synthesis.

The Logical Structure of Meaning

Bivalent action synthesis

Paul beats Peter   or: Paul gives Peter a beating
Peter gets a beating from Paul   or: Peter is beaten by Paul

Obviously, semantic inversion changing the points of reference, is independent of *informational shifting*. The novum (that is what the speaker considers a new information not yet know to the listener) may still be represented by either of the two semantically inverted substances - according to whether we stress Paul or Peter (novum underlined):

Paul is Novum
a) <u>Paul</u> beat Peter   or: <u>Paul</u> gave Peter a beating
b) Peter got a beating from <u>Paul</u>   or: Peter was beaten by <u>Paul</u>

Peter is Novum
c) Paul beat <u>Peter</u>   or: Paul gave <u>Peter</u> a beating
d) <u>Peter</u> got a beating from Paul   or: <u>Peter</u> was beaten by Paul

Both a and b answer the same question: But who was the guy who beat Peter? Or, in c and d: Who was the guy who beat Paul? Novum and Topic thus respond to an informational requirement that is independent from semantic inversion - in traditional Western Grammars commonly called Passive Voice.

Even *trivalent action syntheses* permit shifting the focus. In other words, we may define the semantic relationship existing between them in three different ways. Take the following example:

Trivalent action synthesis

a) **Paul** gives Peter the present

b) **Peter** receives (gets) the present from Paul or:
   Peter is given the present by Paul

c) The **present** is given to Paul by Peter or:
   The **present** is received by Peter from Paul

The *Possession Synthesis* too allows semantic inversion:

The Logical Structure of Meaning

a) **Paul** owns the house (Paul is the owner of the house)

b) The **house** belongs to Paul or:
   The **house** is owned by Paul

Note that in the present as in other examples overlaid semantic shades need not be exactly identical. Thus 'belongs to' may, *but need not*, convey the same semantic content as 'is owned by'. This is quite normal. Any language disposing of more than one expression for certain semantic contents will try to differentiate between them in whatever so insignificant way. In the present context I merely want to stress the fact that semantic inversion is a purely semantic phenomenon and as such quite independent of possible formal realizations.

In the possession synthesis the terms concerned have a similar logical status. This is still the case in the 'localizing synthesis' *with roughly equal terms*. Thus, we may say:

Localizing synthesis

a) the **ball** is in the box
b) the **box** contains the ball.

However, inversion is hardly possible as soon as one of the terms is extremely large and immovable, while the other is characterized in the opposite way. We may easily say 'the ball is in the garden' where the focus is placed on the term which is small and movable in relation. However, we will not say 'The garden contains (bears, has, etc.) the ball'*.

In other words, the process of mentally analyzing reality defines a spatial relationship between two objects in such a way that it focuses the relatively small and movable, which may appear in *different locations so that information can choose between possible alternatives*. Big objects allow for a similar choice only when they become small with regard to still bigger entities. Thus, we will not semantically invert the synthesis 'a comet fell on the earth' by saying something like *'The earth received the comet'* (unless this is done with specific poetic intention), but we will say 'The earth approached Venus', etc. We will see later that to a certain degree the requirements of information determine the possible choices of semantic inversion.

The Logical Structure of Meaning

5 Connections and Conjunctions

Both connections and conjunctions are made of more than one (nuclear or enlarged) synthesis: they are based on a structure of meaning wider than the single synthesis. I call *connections* the combination of the free with a bound synthesis, while *conjunctions* result from the combination of several types of synthesis in their free state. Consider the two following examples:

a) All people fleeing their countries have been accepted as guests
b) People flee their countries because of political harassment

Both express more than one synthesis but only the second is a conjunction, where two syntheses in free state come together. We could likewise say 'People flee their homes. The Reason: They are politically harassed'. The first is an instance of what I call 'connection'. Examples of connections are:

Small dogs like biting
The hat on the table belongs to Paul
The man walking along the road is a stranger
The man, who walks along the road, is a stranger

Connections result from the needs of information. While a free synthesis like 'These (or there) are small dogs' conveys information, the corresponding bound synthesis 'small dogs' does not (see below, III,4). The speaker supposes the listener to know that there are 'small dogs'. Bound syntheses are nothing more than further specified substances, they do not change the logical status of the synthesis concerned.

This is what distinguishes them from conjunctions. The analysis of reality by the human mind, first, leads to a definite range of 'Types of Synthesis' found in all developed languages; and it leads, furthermore, *to a definite range of possible logical relations between them* -- these are what I call 'Conjunctions'. So, these belong to the logical and not to the informational structure of meaning. The more frequent among these are

a) **Psychic-State Conjunction**:
He knows that the rain will come soon.

b) **Time-Space Conjunction:**
After dinner (after they had dined) they came to join us.
During my absence visitors were flowing in.

The Logical Structure of Meaning

They arrived at the place where the others were waiting for them.

c) Causal (Logical) Conjunction:
We were happy because swallows at last arrived.
They came to see us though the road was bad.

The Psychic-State Conjunction links a subjective synthesis to any other that may be either objective or subjective. Consider the following examples:

subjective synthesis	**objective synthesis**
He, know, X	X = rain,come,soon
	He knows that the rain will come soon
subjective synthesis	**subjective synthesis**
He, know, X	X = she,feel,bad
	He knows that she feels quite bad

Some types of subjective synthesis may only appear in combination. While we may say 'I feel well', which is a complete subjective synthesis, it is not possible to say 'I foresee'*. This type of synthesis is incomplete without the semantic content (the synthesis) which is the object of such foresight.

Psychic-State Conjunctions appear in three main semantic variants:

perception synthesis:	He, see, notice, perceive, etc.., X (X = Ship, arrive)
	He noticed that the ship arrived
apperception synthesis:	He, know, presume, doubt, suspect, etc., X (X = ")
	He suspects that the ship will arrive soon
affection synthesis:	He, like, want, be afraid of, regret, etc., X (X = ")
	He wants the ship to arrive soon

In form, the Psychic-State Conjunction is susceptible of widely varying formal realizations not only in different languages but even within one and the same. Let me illustrate this point by stressing the more radical alternatives:

The Logical Structure of Meaning

Different Formal Realizations of semantically identical conjunctions

Apperception synthesis: We (everybody, one), know, X
X = the ship comes
One knows (it is known) that the ship will come
The ship will certainly come
We (everybody, one), assume, X
It is assumed that the ship comes
Presumably (probably), the ship will come
Affection synthesis: Everybody (one etc.), hope, X
X = he comes
It is hoped that he comes
If only he came!
Hopefully, he will come
Everybody (one etc.), regret, X
I am afraid he comes.
German: Leider kommt er.

Expressions like English 'certainly', 'presumably', 'alas', 'hopefully', etc. leave the subjective agent unexpressed. The latter may simply be identical with the speaker or may be a definite, generalized one (everybody, one, etc.). In both cases the zero-form for realizing the agent is used as a means of formal expression (cf. 'Semantic Effacement'). Examples like 'Probably, the ship will arrive', 'They will certainly surrender', 'Alas, they failed', etc. belong to this type of formal realization.

If the agents in the subjective and the objective synthesis are identical, English - and many other languages - resort to formal abbreviation (Formal Ellipsis).

I decided **that I** would go	=	I decided **to** go
You decided **that you** would go	=	You decided **to** go
He decided **that he** would go	=	He decided **to** go
etc.	=	for identical agents

In these examples the formal element 'to' is not a dummy formant as it represents an orderly set of semantic contents (here represented in form by: 'that I', 'that you', 'that he', etc.). What is usually called the 'infinitive' is an action formally realized without a formally realized agent.

The Logical Structure of Meaning

Chomsky (1957:94,100) seems to assume that language may carry dead elements without meaning. According to him, the formant 'to' in utterances like 'he wants to go' is bereft of meaning. Lyons (1987:158) argues in a similar but more cautious way. This argument would imply that language may, in principle, contain merely physical elements without mental content. It is easy to show that this assumption is wrong. If indeed the formant 'to' were a dead element, then any other formant or no formant at all (zero-form) could take its place. Instead of 'he wants to go' it should be possible to say 'he wants rem go', etc. This is clearly impossible and for that simple reason the positions of Chomsky and Lyons are quite untenable. Their views are the result of an unduly narrow understanding of meaning. It is by no means true that the element 'to' is without meaning. Its meaning is to provide a formal abbreviation for linking the two syntheses of a conjunction.

In a number of languages bound formants (mostly suffixes) are used for the purpose of expressing wishes. This is the case in the Sanskrit desiderative:

pipâsâmi = I want to drink = I want (that) I drink
pipâsasi = You want to drink
etc.

Japanese too realizes this semantic pattern with bound formants but the agent may remain unexpressed in form:

nomi-*tai* = (I, you, he etc.) *want to* drink

The **Time-Space Conjunction** creates definite relations between two or more types of synthesis but the semantic nexus need not be reflected in form. A single English sentence may serve to realize a time-space conjunction, but it may be realized in two sentences a well.

a) We went home at sunrise
b) We went home when the sun was rising
c) The sun rose. During that time, we went home

In all three cases, the logical structure of meaning is identical though formal realization is quite different. In the first case, traditional grammar speaks of a temporal specification, in the second of a 'subclause', while (c) represents in form the two independent semantic syntheses. It goes without saying that the semantic identity in question only obtains in view of the logical structure

The Logical Structure of Meaning

of meaning. As far as style is concerned, such formal alternatives are of great importance. Style relies on subtly superimposed semantic shades as well as on playing with form; there is never complete identity. The fascination of language is largely due to the fact that above and beyond the logical structure of meaning and its alternatives of formal realization, style plays with the more sophisticated aspects of linguistic expression. This, however, is not what I am speaking about in this book. From the vantage point of our analysis, the three formal variants realize an identical semantic conjunction.

Causal (Logical) conjunction are especially rich in formal alternatives. The causal relationship may, for instance, lead to the following equivalents:

a) The shop is closed because (though) I am ill
b) The shop is closed due to (despite) my illness
c) I am ill. Therefore (Nonetheless) the shop remains closed

Scholars of Sanskrit know that, especially during the later phases of that language, a single case (Sanskrit ablative) was used to express the causal relationship. Here an example from the Mahabharata:

yasya dandabhayât sarve dharmam anurudhyanti
whose rod-fear-from all duty follow
= from fear of whose rod all are constant to duty (Whitney, 97)

Some conjunctions are based on a specific logical relationship plus a specific type of synthesis, as, for instance, a causal relationship combined with an affection synthesis. Expressions that are realized by means of 'in order to' presuppose such specific combination:

We went there in order to help him
We went there because we wish to help him

Semantically, this expression contains three different types of synthesis:

1) We go there, Causal nexus: 2) We want X   3) X = We help him

III The Informational Structure of Meaning

The elementary Types of Synthesis belong to what I called the ‚*Logical Structure of Meaning*' (logic in the larger linguistic sense). Of equal importance is its counterpart the ‚*Informational Structure of Meaning'*, as it is the latter that defines the *purpose* of language. By themselves semants and the different types of synthesis to which they give rise, are but reflections of reality (according to processes of the human mind). They show how the human mind, after analyzing reality into concepts, recomposes it into definite semantic patterns. These patterns are quite independent from individual concepts. It is exactly for this reason that I speak of 'types'. A mere duplication of outward reality in the minds of human beings would be a superfluous luxury if it would not either enhance our knowledge of surrounding reality or be used as an instrument for action. That is precisely what the *informational* devices of language achieve.

The informational Structure of Meaning consists of the following seven elementary parts:

1 Commands (Requests)
2 Statement / Question
3 Topic / Novum
4 Free / bound Synthesis
5 Semantic explicitness / semantic effacement
6 Total effacement and rank lifting
7 Derivative us of semantic effacement
8 Semantic tingeing

The Informational Structure of Meaning

1 Commands (Requests)

The oldest parts of any natural language are most probably made up of commands as is still true of animal languages. Flee! (there is a tiger around). (Please) come here! (there is something to eat). Be careful! (or you fall down the cliff). Up to the present day, this explains why languages generally choose the shortest possible way of formal realization like go!, hit!, climb!, eat! Japanese uses very elaborate forms when the command is meant as a mere exhortation or recommendation. Then it uses formulas like 'be so kind as to go' (itte kudasai) but it is as succinct as any other language when the command is meant as a true command: 'ike!' (go!), 'koi!' (come!).

A mother speaking to her child will start language with similar exhortations: drink!, stop drinking! and so on. It will take some time before she makes descriptions 'See how beautiful the balloon' or 'Now we take you in the room with air-conditioning'. Descriptions are therefore likely to come later than the most primitive linguistic stage made of commands or - in the case of infants - affectionate exhortations.

Commands express a forceful wish - for this reason they represent a special case of the Psychic-State Conjunction:

I, want, X   I want that you go. I want you to go. Go!

Gradually human beings acquire a more complex knowledge of reality - they learn to analyze its structure creating the types of synthesis as means of description. But description is no aim as such it serves to convey information between speakers and persons addressed. Such a traffic of information occurs in the shape of

2 Statements and Questions

Most *statements* imply a transfer of knowledge from a speaker to a listener. The first would not utter a sentence like 'The tower has a height of 200 feet' if he believed that the listener already knows the fact. In case he himself wants to be informed, he turns the synthesis into a *question*: "What is the height of this tower?"

The Informational Structure of Meaning

Any free synthesis may take either the semantic appearance of a statement or that of a question. This difference does not concern its logical structure, it exclusively concerns its *informational* use.

In form many languages treat statements, questions and commands in a similar way: 'He goes home', 'Does he go home?', 'You go home!' This has induced linguists in their descriptions of language to deal with them as if they belonged together. But while question and statement are true informational variants of logically identical units, commands imply a fundamental change on the level of logical structure. Statements and questions presuppose a description of reality, commands are meant to induce action. For this reason, every command is - as stated before - a conjunction.

It may be argued that the breakthrough to human language results from the existence of questions. When a speaker asks the addressed person: "Where are you going?" Or: "Where did you get your wound?", the borders of the here and now are crossed in both instances. The answer must relate to future or past events, which is not yet the case when the emitter of signals and their receiver both remain enclosed in the here and now (at most one of them enjoys a broader view so that he can issue warning calls in case of imminent danger, but the threshold leading forward into the future or backwards into the past is not exceeded).

So, it is indeed the existence of questions that provide us with the distinguishing feature of human language when compared to its non-human antecedents (in the beehive the returning bee is not questioned, but it is genetically programmed to pass on the location of a new food source with reference to the hive, and it cannot, of course, answer any other questions beyond the genetically fixed).

Even questions do not, however, fall from the sky - when viewed from an evolutionary point of view. Dog owners know quite well that their pets often attempt to induce them to play. These attempts are nothing else than elementary questions, whose answer in the way of consent or rejection is quite well understood by the animal. But such elementary antecedents are again limited to the expression of needs relating to moments and situations in the here and now. Only the special achievement of humans to *ask for informations beyond the here and now or the immediately visible has caused the extraordinary explosion of memes* (the cultural counterpart to genetic endowment). And that certainly constitutes an evolutionary breakthrough of unique importance, because the knowledge of each individual *can be multiplied almost ad infinitum.*

3 Topic versus Novum

The logical structure of meaning is at the base of its informational counterpart, and it remains unaffected by any use it is put to by the requirements of information. But when the logical structure of meaning is put to informational use, any speaker addressing a listener bases his utterances on assumptions as to what the latter already knows or does not know.

Consider the case of a man running onto the street and crying. <u>'The house down the road is on fire.</u>' Supposed that people having just arrived from outside have so far neither seen the house in question nor the fire down the road, then the information *as a whole* is new to everybody - in other words, it constitutes the Novum. It could, therefore, be expressed by two independent syntheses as well: '<u>There is a house down the road.</u>' '<u>It is burning.</u>'

But now let us suppose that everybody has already seen the fire and somebody starts crying 'The fire <u>is getting out of control</u>!' The information conveyed in this case is quite different. One part of the utterance - the fire - refers to something already known, that is the Topic, only the remaining part represents what is informationally new. While in the first case the entire utterance constituted the Novum, that is what the speaker supposes to be unknown to the listener, the utterance is now made of two parts: The Topic and the Novum.

Traditional linguists have spoken of 'Topic' and 'Comment' but this distinction is not wholly adequate as it leaves no room for the case that an *utterance as a whole* represents what is 'informationally new'. Of course, no utterance can be a Topic as a whole - this would contradict the very purpose of information, but it can very well be an informational novum as a whole. *This possibility excludes a further identification - that of topic and novum with subject and predicate.*

In German you may find expressions like '<u>man lacht</u>', '<u>es wird gelacht</u>' (<u>all people are laughing, everybody is laughing</u>). Here the whole synthesis is meant to be a novum, so it never answers questions like 'Who is laughing?' It can only answer questions like 'What is happening?' Obviously, what traditional grammar here designates as the subject, namely 'man' or the dummy formant 'es', do not represent the topic.

It may, however, be argued that in some exceptional cases utterances may as a whole figure as topics. Suppose that somebody remarks 'it is cold in this room' and somebody else repeats: 'yes, indeed, it is cold over here.' The repetition could be viewed as an informational topic comprising the whole utterance because it does not convey any new information. Indeed, if it were nothing but a mere repetition it would make no sense from the standpoint of information. But, normally, such utterances imply *a kind of confirmation* because the listener might after all disagree. The fact that the listener *confirms* the assessment made by the speaker represents the informational novum in this as well as in similar cases.

Any synthesis containing more than one part, for instance, more than one substance is susceptible of *informational shifting* – either Novum or Topic may be represented by any substance. Thus, in a bivalent action synthesis like 'Peter(Ag) beat Paul(Pt)' it depends on stress whether Peter or Paul is understood as Novum (and the other person as Topic). Some languages - among them English - prefer the Novum in the head position of a sentence. If Paul(Pt) is meant to be the Novum (<u>whom</u> did Peter(Ag) beat?), then a different construction like '<u>Paul(Pt)</u> was beaten by Peter(Ag)' is more appropriate to fulfil this condition. That is why informational shifting may go hand in hand with *'semantic inversion'* (see II,4), where the logical point of reference has been changed. Consider the following example of a bivalent action synthesis.

Bivalent action synthesis   **Paul(Ag), Peter(Pt), beat**

a) the Agent as Novum: <u>Peter</u> beats Paul   (<u>who</u> beats Paul?)
b) the Patient as Novum   Peter beats <u>Paul</u>   (<u>whom</u> does Peter beat? Paul being stressed)

Stress is, of course, only applicable in the spoken language, so the difference between Topic and Novum would not be recognized in a written sentence. This difficulty is removed by way of *semantic inversion* where the head position is assigned to the patient, so that it may assume the informational role of Novum:

a) Peter beat <u>Paul</u> (stressed in spoken language)

The Informational Structure of Meaning

becomes through semantic inversion in written language:

b) <u>Paul</u> was beaten by Peter   or:

c) <u>Paul</u> got a beating from Peter

All three examples answer the same question: <u>Whom</u> did Peter beat? However, in spoken language the head position of b) and c) may still be *overruled* by stress so that Peter may still become the Novum.

A relational quality synthesis containing more than one substance is susceptible of informational shifting just like an action synthesis. Note that the distinction of actions from qualities is equivalent to that of dynamic versus static properties. It should, therefore, not come as a surprise that 'semantic inversion' is a common property of the quality synthesis as well, so that the Novum may, here again, be placed at the head of sentences.

<u>the tree</u> is bigger than the wall   <u>the wall</u> is smaller than the tree
<u>the tree</u> surpasses the wall*   <u>the wall</u> underpasses the tree*
<u>the tree</u> is underpassed by the wall*   <u>the wall</u> is surpassed by the tree*

In English the verb 'surpass' is normally used figuratively but in German it is perfectly correct to say 'der Baum überragt die Mauer' and 'die Mauer wird vom Baum überragt'. Examples like these clearly show that *the so-called passive voice is no more than one among other formal means to express semantic inversion*. Semantically all examples to the left define the relation between tree and wall with reference to the tree. After semantic inversion all examples to the right define it with regard to the wall. As in English, the Novum tends to be placed at the beginning either tree (left) or wall (right) are normally assigned the role of Novum. As we have seen, this rule may be invalidated by stress in spoken language.

Let me add that any non-action synthesis containing more than one substance may likewise be subject to informational shifting. As example consider the following 'localizing synthesis':

<u>Lots of people</u> are in the house   <u>the house</u> is full of people

The first example may come as an answer to the question: Is there <u>anybody</u> in the house? The second: <u>Where</u> are all those people?

The Informational Structure of Meaning

Informational shifting not only occurs between substances as illustrated above, it may extend to any semant within a synthesis. Formal *lifting of rank* (in those languages where it is possible at all, see II,5) then comes to play a prominent role.

Let me illustrate informational shifting with regard to the following simple action synthesis:

I	II	IV	III
lion	roar	very	noisily

Let us assume this utterance to occur within a novel where there has been no previous reference to lions or noises. Then the utterance as a whole represents the 'Novum'.

Now suppose some person within the novel may want to refer to any part of this statement, making his comments on it and thus offering an informational Novum.

	I	II	IV	III	I (new first rank position)
a)	Lions	roar	very	noisily	
b)	<u>Lions</u>	roar	very	noisily	
c)	Lions	<u>roar</u>	very	noisily	The <u>roaring</u> of lions is very noisy
d)	Lions	roar	very	<u>noisily</u>	The <u>noisiness</u> of leonine roaring (is amazing)
e)	Lions	roar	<u>very</u>	noisily	The <u>degree</u> of noisiness (is amazing)

Informational shifting may thus extend to all four semants according to the question to be answered.

a) What happens   Lions roar very noisily
b) <u>What</u> are those roaring animals?   <u>Lions</u>.
c) <u>What is remarkable</u> about them?   Their <u>roaring</u>.
d) <u>What is the quality</u> or their roaring?   (Unbearable) <u>noisiness</u>.
e) <u>What do you find so astonishing</u> in their noisy roaring?   The <u>extent</u> of their roaring

English is a highly developed language that permits rank-lifting; it is therefore quite easy to single out any semant and make it appear in a first rank position as Novum.

Languages without the formal means of rank-lifting are equally capable of conveying the information the speaker wants to transmit. Instead of saying '<u>such noisiness</u> is too much for me', the speaker may, for instance, say 'yes, too <u>noisy</u> for me'. Or, in a more detailed manner, 'yes, the lions roar <u>noisily</u>, much too <u>noisily</u> for me.' Here, as in the previous example, the informational Novum is contained in the respective evaluation of the noise. Such examples illustrate that English informational structure has no advantage over others but it offers *more than one* formal alternative of realization. In informational shifting, rank-lifting is not a necessary formal device but it offers an easy formal solution.

Formal ellipsis, that is, realization as zero-form, is frequently used for informational Topic (that is for what is already known to both the speaker and the listener). It could be a standing device if established formal patterns did not contradict it. Formal realization in English admits sequences of question and answer like the following:

Question: <u>Whom(Pt)</u> did Doreen marry? Answer: <u>Robert(Pt)</u>

The complete form: 'She(Ag) married Robert(Pt)' is not obligatory in English. The same does not hold true in other cases.

Question: <u>What</u> did Mary(Ag) do? Answer: <u>Walked home</u>*

Here, English does not permit formal ellipsis though 'Mary(Ag)' (she) - as the part representing the Topic - is known to both speaker and listener, and its formal repetition is therefore just as redundant as would be the corresponding repetition in the first example. It is the established formal pattern of synthesis realization, which, in similar cases, does not permit English to omit the agent. The correct answer in English must be:

Question: <u>What</u> did Mary(Ag) do? Answer: She(Ag) <u>walked home.</u>

Other languages like Japanese make extensive use of formal ellipsis in this and in similar cases. For instance, somebody seeing a child (Topic) bent over a table may ask '<u>taberu</u> no?' (= <u>eating</u>?). No formal element expresses the agent but from the *situational context* it is perfectly clear to both the speaker and the child addressed by him that only the latter may be referred to. This, too, is a kind of formal abbreviation, but this time it is not based on the fact that the semantic content has been previously expressed in form. Instead of being based on formal context it is based on a situational one.

The Informational Structure of Meaning

The use of form in language is to make manifest by means of physical substitutes for mental events what otherwise would not be understood. If the semantic content in question is understood anyway (because of previous formal or of accompanying situational context), form has no indispensable function to fulfill. Without formal realization both parties are perfectly aware of what is meant. In a general but incomplete way[8] this explains why, in Japanese, the Topic (in traditional grammar the 'subject') will be frequently omitted. More often languages do, however, take an opposite course: They are redundant - their formal systems oblige speakers to express contents even if these are obvious from the context.

In informational shifting, as in semantics at large, formal realization must express what would otherwise give rise to misleading interpretations. It must be clear whether the speaker is talking about something already spoken about in the course of conversation, or whether he is introducing a new subject and thus provides new information. But languages differ in that some make extensive use of formally expressing what from a logical point of view need not be expressed. For instance, many languages use obligatory time specification in every sentence (tense) although such systematic use is redundant in most cases; in the same way there are others that resort to a rigorous distinction of Topic versus Novum beyond logical necessity.

4 Free versus bound synthesis

In the following chapter I will discuss what in Chomskyan grammar appears under the heading of Recursion or Embedding. *Both are purely formal, in other words strictly "meaning-less" concepts. Any formal elements may indeed be recurring or be embedded.* These terms are without value in Universal Grammar.

In his book "The Language Instinct", Pinker defines recursion as follows: "Recall that all you need for recursion is an ability to embed a noun phrase

---

[8] The explanation covers all cases but is not wholly sufficient. While it explains the possibility of the formal device in question, it does not say anything about the motivations which may have prompted its use. Japanese don't like to stress their Ego, so formal ellipsis of first person subjects has an important socio-cultural function.

The Informational Structure of Meaning

inside another noun phrase or a clause within a clause... With this ability a speaker can pick out an object to an arbitrarily fine level of precision". Here Pinker takes the decisive step by emphasizing precision, **that is meaning**, but he does not further specify the range and nature of meaning made more precise. That is indeed the task of General Grammar.

Let us analyze a simple statement like 'The tree <u>is now green again</u>' where tree represents the Topic while 'is now green again' constitutes the Novum or the information the speaker wants to convey. Now, the Topic is a single semant, a substance in this case, but it may be replaced by a synthesis, in fact by any type of synthesis in a bound state:

Statement containing a /Bound Synthesis/	As Free Synthesis
*/The green tree/ <u>is green</u>	The tree is green
/The old tree/ <u>is now green again</u>	The tree is old
The tree /that we saw a year before/ <u>is now green again</u>	We saw the tree a year ago
/Peter's tree/ <u>is now green again</u>	The tree belongs to Peter
/The tree on top of the hill/ <u>is now green again</u>	The tree is on the hill

What makes the single substance tree similar to 'old tree', 'tree we saw a year ago' etc. is that each represents the Topic that is what the person addressed is supposed to know. The distinction of information believed by the speaker to be known or unknown to the person addressed is indeed of crucial importance. This is seen by the first example, which is strictly tautological since it conveys an information already known to the person addressed.

The distinction of information known and unknown to the person addressed pertains to all types of synthesis. These may therefore appear in two entirely different semantic shapes: either as '*information*' or '*non-information*' like, for instance, the two English Action Syntheses 'Men eat rice' and 'Men eating rise (<u>are usually healthy</u>)'.

Structure of meaning	**Possible English realization**
Action Synthesis: Men(Ag), rice(Pt), eat	
a) Conveying information unknown to the listener	<u>Men eat rice</u>
b) Not conveying new information to the listener	Men eat rice, you know. (<u>They are usually healthy</u>)

or: /Men eating rice/
(are usually healthy)

or: Men /who eat rice/ (are ")

In the first instance a), some fictitious man of the moon may just be explaining the eating habits of terrestrials. The speaker provides true information as he supposes to say something new to the listener.

The second instance 'Man eating rice (are usually healthy)' contains two types of synthesis, first, an Action-Synthesis (Man eating rice), and, second, a Quality-Synthesis (they are usually healthy). Here the Action-Synthesis (men eating rice) does not convey any information as the listener is supposed to know that there are men, who eat rice. Relevant information is only provided by the Quality-Synthesis, which states that these men are invariably healthy. In both cases, the *logical* structure of the Action Synthesis is identical, but its *informational* function is different. English has special formal means to express this difference on the formal plane ('men eating rice' or 'who eat rice'), but, as pointed out by Hallpike,[9] primitive languages do not necessarily dispose of such a ready-made formal shortcut to express the non-information synthesis. Instead they use devices like the above: 'Men eat rice, you know. (They are usually healthy)'.

Let me use a more convenient way to distinguish both functional types of synthesis by calling the Information Synthesis 'free' and the Non-Information Synthesis 'bound', *as the latter must always be part of another synthesis that conveys information.* This statement is impressively confirmed when bound and free synthesis express the same meaning. *Men /eating rice or who eat rice/ are (in the habit of) eating rice. As the novum adds nothing new to the topic, the sentence is tautological though grammatically perfectly correct.

Categories and the bound synthesis

The free synthesis is the smallest unit of information. It was said earlier that the synthesis mirrors the previous analysis in the mental processing of reality

---

[9] http://www.gerojenner.com/wpe/the-hallpike-paper-universal-and-generative-grammar-a-trend-setting-idea-or-a-mental-straitjacket/

The Informational Structure of Meaning

- but in the opposite direction. Categories alone (such as substances, actions, qualities, etc.) are but the elements out of which a mentally reconstituted part of reality is made. For this reason, categories as such do not convey information. The same applies to the bound synthesis which as part of a free one *has the same status as a category*: It does not constitute a unit of information.

Traditional concepts in Comparative Linguistics

The incompleteness of traditional grammar and its often absurd consequences become especially obvious with regard to its treatment of the so-called relative clause.

In a sentence like 'The man, who walks along the road, is my friend', the Topic initiated by the relative pronoun 'who' is said to be a relative clause. Now, the following sentence 'The man walking along the road is my friend' is not initiated by a relative pronoun. As there is no longer any relationship between a relating element (who) and an element it relates to (man), the expression 'walking along the road' cannot possibly figure under this heading. It is, however, quite obvious that both sentences represent mere alternatives of formal realization of identical semantic structures.

On the other hand, the so-called relative clause of traditional grammar is in itself a mixed thing since it may refer to semantically quite different contents. In the two sentences:

1) Mr. Abbot, who by the way is on the way to London, is my personal friend, and:
2) 'the man, who goes to London, is my personal friend'.

the so-called relative clause does not serve the same semantic purpose. In the first instance we may cut the expression into two free syntheses. 'Mr. Abbot - he goes to London right now - is my personal friend', or we may say 'Mr. Abbot is my personal friend, by the way, he goes to London at present'. However, it is quite impossible to split up the second example. We may only use the formal alternative already mentioned above: 'The man going to London at present is my personal friend'.

As a general tool of description the term 'relative clause' proves to be useless for still another reason. There are languages without relative clauses.

The Informational Structure of Meaning

Formal realization in Japanese and Chinese exclusively proceeds according to a pattern similar to English 'the man going to London <u>is...</u> '.

The following *connections* consist of two combined syntheses, a bound and a free one. Their logical status is different in each case. In (1) a specific and definite green tree, in (2) several definite green trees, in (3) and (4) all green trees are said to be big. (3) and (4) represent formal alternatives, both mean that, as a rule, green trees are big. These examples are meant to show that the informational dichotomy of free versus bound remains unaffected by further modifications of the logical structure:

Synthesis: free/bound   **formal realization in English**

"quality synth."; "/quality synth./"

1) "tree, big"   "/tree, green/"   /The green tree/ <u>is big</u>
2)   /The green trees/ <u>are big</u>
3)   /Green trees/ <u>are big</u>
4)   /A green tree/ <u>is big</u>
5)   /One of the green trees/ <u>is big</u>

All examples are identical in that one and the same type of synthesis (quality synthesis) appears in both its semantically free (tree is big) and bound (green tree) informational variants.

It should be noted that in the English Quality Synthesis the position of substance (replaced or not by a bound synthesis) is fixed so that the Novum cannot change place as in the bivalent Action Synthesis where it may come at the head or the end of the utterance. 'The man with the knife <u>hit Peter</u>'. Or: '<u>It was Peter who was hit</u> by the man with the knife'.

The semantic distinction between a free and a bound synthesis is a *universal* trait of natural languages as it is based on equally universal differences concerning the availability or need of information.

The Informational Structure of Meaning

Every synthesis may be either free or bound

Look at the following types of synthesis, which were explained earlier. In the second and the following examples I refer to the respective formal realizations in English.

Free synthesis	**Bound synthesis**
Quality Synthesis: /Tree, green/	/the green tree/... or: the tree /that is green/...
Action Synthesis (Agent/Patient): /Dog(Ag), cat(Pt), bite/	/the dog biting the cat/... or: the dog /that is biting the cat/...
Possession synthesis (Possessor/Possessum): /Peter(Pr), Ball(Pm), belongs/	/the ball belonging to P/ ... or: the ball /that belongs to P/ ... or: /Peter's ball/...
Localizing synthesis (large/small): /House(s), hill(l), stands/	the house /that stands on the hill/... or: /the house on the hill/...

A synthesis will only be functionally bound if, as a free one, *it would convey superfluous information*. On the other hand, it cannot be bound if no superfluous information is conveyed - regardless of whether or not its pattern of formal realization is equal to that of the pattern normally used for the bound synthesis. As a general rule, substances representing proper names cannot be part of a bound synthesis.

As already mentioned, the English sentence 'Mr. Abbot, who is walking home just now, is my friend' follows the same pattern of formal realization as does 'the man, who is walking home just now, is my friend'. But the second example represents a true bound synthesis, while the first may be read in two alternative ways. If the speaker believes that the listener is well aware of the fact that Mr. Abbot is on his way home, then it is a bound synthesis without new information. He might say as well. 'Mr. Abbot who, as you know, is walking home just now, is my friend'. The information is only confirmed without being new and the speaker apologizes for uttering it with the words: 'as you know'. If, however, the speaker wants to say 'By the way, Mr.

The Informational Structure of Meaning

<u>Abbot is just on his way home</u>' then this is new information and both syntheses are semantically free. As a rule, any substance represented by a proper name must be known to the listener. For that reason, *a proper name is normally not in need of further specification by a bound synthesis.*

It is one of the main defects of traditional grammar to draw its boundaries according to outward differences of form, thus obliterating underlying identical semantic patterns or, on the contrary, overlooking semantic divergencies because they do not appear in form. The bound synthesis allows for quite a range of different formal realizations not merely between languages but even within one and the same language.

For instance, /Peter' s hat/ and /the hat belonging to Peter/ may not be the same to a poet but, as far as the difference between free and bound synthesis is concerned, they are as strictly equivalent as are 'the green tree' and 'the tree /that is green/'. The same applies to formal alternatives like 'the book /that belongs to me/' and '/my book/' or (in a localizing synthesis) 'the tree /that stands on the hill/' and '/the tree on the hill/'. By providing quite different descriptions to clauses like 'the tree on the hill' and 'Peter's hat', traditional grammar failed to grasp the underlying semantic base.

We saw that any bound synthesis represents nothing more than a specification of the category substance. Such specification does as a rule not go beyond the level of a single synthesis like in '/the man going to town/ <u>is my friend</u>' , but in literary language it may well go much farther. In a sentence like 'the man /who is coming while smoking his cigarette and shaking an umbrella over his head/ <u>is my friend Peter</u>', two bound syntheses follow each other. We would again create unwanted information if we changed them into their free counterparts. The only true information is provided by the free synthesis: <u>He is my friend</u> - all the rest is Topic, that is known to both the speaker and the person addressed.

It is, however, quite rare (at least in spoken language) that bound expressions exceed the limits of a synthesis. Normally a rough sketch of the situation will sufficiently guide the attention of the listener.

In General Grammar, the distinction between the free and the bound synthesis is strictly defined, and this definition *exclusively* relies on semantic criteria. However, the use of form to represent this distinction is often equivocal. Consider the following examples:

The Informational Structure of Meaning

They did /the quick job/   or: the job /that was quick/
They did a quick job   or: they worked quickly

In the first instance, the listener knows that there are several jobs, one of them being of short duration, so /quick job/ is a bound synthesis. The second case represents a free action synthesis modified by the second rank specification "quick". Any part of both utterances figure as Novum or Topic.

General pattern of formally realizing the bound synthesis

I will first use the traditional concepts S, O and V to illustrate formal realization of both the free and the bound synthesis. Next, I will show why they are inadequate, as they were originally derived from classical European languages Greek and Latin. Indeed, the use of these concepts not only complicates understanding but makes it both impossible and contradictory.

I) The easiest way to distinguish S and O – and the one characteristic of Latin, Greek, Sanskrit and most other languages – is designation.

S and O are distinguished by *adposition or case*, usually the nominative in the first, the accusative in the second instance (but in Russian the genitive is likewised used for this purpose). As the persons reading this book are familiar with the English language, I will first introduce relevant English examples, though these *do not* distinguish subject and object by designation but by position (word order).

    Let me illustrate the basis semantic structure that I want to explore by the following English sentences constisting of a free together with a bound synthesis:

a1) **The man,** who beat Peter, **broke the stick**
b1) **The man,** whom Peter had beaten, **broke the stick**

The two sentences express a semantic deep structure that consists of two semantic dichotomies:

The Informational Structure of Meaning

1) agent (man, who)/ Peter (object1); agent (man) stick (object2)
with the roles of agent and patient exchanged in the bound synthesis
2) **main clause** / relative clause

Traditional grammar including that of Chomsky uses concepts like S, O and V to describe different languages by means of identical terms. Our two examples therefore appear as follows:

a1) **The man**   who   beat Peter   **broke the stick**
     Subject rpron   V   Object   V   **Object**

b1) **The man** whom Peter (had) beaten **broke the stick**
     Subject rpron   Subject   V   V   **Object**

These terms originally derived from classical European languages like Greek and Latin are inadequate as their use not only complicates understanding but makes it both impossible and contradictory. They don't even allow us to understand the two following logically identical examples *within the same language*, that is, English:

a2) **The man,** by whom Peter was beaten, ...
      S   ? rpron   S   ?   V   , ...

b2) **The man,** who was beaten by Peter , ...
      S   rpron ?   V   ?   ?

Both a1) and a2) and b1) and b2) are stictly identical in logical structure but the elements appearing as questions marks can no longer be accounted for by the terms S, O, and V. And what is still more intriguing: the object of the bound synthesis has all but disappeared – in other words, S and O cannot belong to the deep semantic structure even in one and the same language like English. Indeed, this observation applies to the main clause too:

The man broke the stick
     S   V   O

The Informational Structure of Meaning

The stick was broken by the man
    S   ?   V   ?   ?

Again, both sentences are *strictly identical* as to their logical structure, but they differ in terms of informational meaning. Man is the topic in the first, stick in the second example. While the first example is perfectly described by "S V O", the second instance, *though consisting of identical semantic elements*, can no longer be explained by these terms. Subject and object no longer serve us. We are compelled to look for different terms that can and must be defined in a purely semantic way. These are agent and patient. Both are present in the second instance as well as in the first. They - and they only - are part of the semantic deep structure. We must therefore adopt the following scheme:

The man broke the stick
    Agent   a   Patient

The stick was brok-en by the man
    Patient   ?   a   ?   ?   Agent

The reason why I replace "verb" with "action" (a) is explained in my work *The Principles of Language*. It is of no particular importance in the present context.

Two formal elements still remain unexplained, they are the result of shifting the topic. Man is topic in the first, stick in the second instance. The so-called passive voice here serves as a formal means to achieve topic shifting.

The two examples thus illustrate a common semantic structure consisting of two semantic oppositions:

1) agent/ patient
2) agent in the role of topic/ patient in the role of topic

The man broke the stick
    Agent   a   Patient

The Informational Structure of Meaning

The stick was brok-en   by   the man
Patient x_1 a   -x_2 DsAg   Agent

Now, we fully reveal the semantic elements expressed, that is agent, patient, x for topic shifting and DsAg, the formal element here used for designating the agent. Our description is based on the semantic deep structure and applies to two intra-lingual examples that cannot be described by means of the traditional terms S and O.

Once we accept the common semantic deep structure as the true and, indeed, *the only tertium comparationis within and between languages*, there is no other way than to turn to a purely semantic description. Such a change had already been advocated by the great Danish linguist Otto Jespersen and was renewed by Steven Pinker's notion of "mentalese".

Pinker says: „Mentalese: The hypothetical 'language of thought,' or representation of concepts and propositions in the brain in which ideas, including the meanings of words and sentences, are couched." And: *„Knowing a language, then, is knowing how to translate Mentalese into strings of words and vice versa".*

This is a cogent argument, even though the term mentalese is a misnomer, since the process of forming word sequences at the formal level is a mental phenomenon too. What Pinker should have said is: Knowing a language is knowing how to translate a structure of meaning into a structure of form (strings of words etc.) and vice versa.

The method to be used when we analyze how meaning gets translated into form, is quite precise. In the *Principles*, I developed this method calling it "Differential Analysis".

Differential analysis is an exact method for mapping the semantic deep structure of any language on its formal surface. Insofar as the chosen deep structure is common to several or all languages, it represents the only scientifically sound method of comparative linguistics.

It proceeds in four clearly defined steps. First, determine the semantic structure. In the following, this will consist of three basic distinctions: a) agent/ patient, b) free/ bound synthesis, c) agent in the role of topic/ patient in the role of topic. If all three oppositions are to be formally expressed, the minimum number of formal examples will be 2 x 2 x 2 = 8. But in English the actual number is at least twice as high because of purely formal alternatives (man beating Peter = man who beats Peter) etc.

Second step: take sentences that differ *by just one* of their semantic contents (semants) but are identical as to the others.

Third step: replace the formants that realize this difference with that semant (semantic content) and repeat this for all semants (in our case for all three dichtomies).

Fourth step: for each formant show all semants arrived at in step three. You will then discover that one formant may simultaneously or alternatively realize more than one semant. Some of these semants will be active in one context and suppressed in another, some will be realized synchronously with others, and some not. In this way we get an exact knowledge of which formants realize which semants. In other words, we have uncovered the hidden deep structure of meaning as it appears on the surface, and we understand the complex workings of the human brain. But never forget: *The ultimate aim of replacing arbitrary formants with non-arbitrary semants is to allow intra- and interlingual comparison.*

Differential analysis is not a mere addition to the procedures of traditional grammar - it revolutionizes its very foundation, since it makes sense only if based on pure meaning - the rock on which all formal realization is built. From a logical point of view, it would be circular to ask how formal or hybrid terms (i.e., terms defined partly by form and partly semantically) map onto the formal surface of a given language. This was shown above for the hybrid terms S and O. In the following, it is shown for the hybrid terms main/relative clause, passive, etc. These terms must therefore be replaced by purely semantic ones.

The above-mentioned examples illustrate the two semantic dichotomies of, first, agent/ patient, second, topic shifting, which must be realized by at least 2 x 2, that is four examples (two for the main clause, and two for the relative one).

But let us now again discuss the validity or not of traditional terms. Are main clause and relative clause concepts that can be used in intra- and interlingual comparisons?

Look at the following sentences:

The Informational Structure of Meaning

1a) The gasping fellow left a bad impression.
1b) The fellow, who was gasping, left a bad impression.

and

2a) The green tree stood at the top of the hill.
2b) The tree that is (or was) green stood at the top of the hill.

Traditional terminology sees relative clauses only in 1b) and 2b). But the two examples listed under 1) and 2) are identical both in their logical and informational content. The gasping fellow (who was gasping) and the green tree (that was green) represent the non-informational alternative to their informational counterpart, that is to:

The fellow is gasping and The tree is green.

The shortcoming of the traditional terms, main versus relative clause, should be obvious. They describe the surface structure but not the semantic dichotomy at its base. So far, this deep structure dichotomy had no name. I speak of information versus non-information synthesis or free/ bound synthesis.
   Replacing surface terms with true semantic ones, we now get three purely semantic dichotomies, with which to describe the very core of language:

1) agent/ patient
2) free/ bound synthesis.
3) agent of the bound synthesis in the role of topic versus patient in the role of topic (active/ passive voice)

The first and second semantic alternative must be realized even in the most primitive language, the third is formally realized in all modern developed ones. As stated before, the minimum number of formal examples will be 2 x 2 x 2 = 8, if all three oppositions are to be formally expressed. But in English the actual number is at least twice as high because of purely formal alternatives (man beating Peter = man who beats Peter) etc.
   I will now show how English and Chinese realize this basic semantic deep structure (but I simplify the results of differential analysis so that they can

be more easily understood. The reader will find a more fully developed example of differential analysis at the end of this book.

A comparison of English and Chinese

English a)

The man /beat-ing Peter/ **broke the stick**
    Ag / a - // Pt /   a   **Pt**

The difference of Agent/ Patient is expressed by word order. That of free versus bound synthesis by the formant "-ing", which realizes the bound synthesis.

Chinese a)

/Ouda BIDE de / **ren daduan le gunzi**
/Beat Peter // / **Ag** **broke**   **stick**
/ a   Pt // / **Ag**   **a**   **Pt**

The difference of Agent/ Patient is expressed by word order. That of free versus bound synthesis by the formant "de", which realizes the bound synthesis.

As designation (case or adposition) does not indicate the roles of agent and patient, the sequence "S /V O.." in English a) could be understood as a free synthesis (main clause): "The man beat Peter". That is why there **must be** a formal element "//" indicating the bound synthesis. In English this formal element is "-ing", in Chinese it is "de".

Let me omit the logical alternatives of a): The man who was beating Peter and a2) The man by whom Peter was beaten.

I will now turn to the logical alternative b) in English and Chinese.

English b)

The man / Peter   has beaten/...
    Ag / (//)Ag   a /...

English formally realizes the bound synthesis by placing its agent immediately after the agent of the free synthesis, that is, by means of word order. This is indicated by putting // in brackets. This possibility does not exist in Chinese as the bound synthesis precedes the free one:
Chinese b)

/Bei   BIDE ouda   de /   **ren...**
/ by   Peter beat   // /   **man...**
/ x   Ag   a   // /   **Ag...**

Chinese must use the passive voice, that is the formal device for shifting the topic (expressed by the semant "x") in order to formally realize the second logical alternative. For the sake of easier understanding, I have translated Chinese "bei" with English "by" but it really transforms the verb into the passive voice.

General scheme of formal realization:

Alternative a)

English: Ag   /a-// Pt /   **a Pt**

Chinese: /a Pt // /   **Ag a Pt**

Alternative b)

English: Ag / (//)Ag a /...

Chinese: /x Ag a /// **Ag...**

In both languages, word order is maintained in the bound synthesis too (a Pt in English and Ag a in Chinese).

Once more let me emphasize what differential analysis is all about. *It reveals the semantic deep structure within and between languages. The*

The Informational Structure of Meaning

ultimate aim of replacing arbitrary formants with non-arbitrary semants is to allow intra- and interlingual comparison.

Not all languages formally realize the bound synthesis

In theory every type of synthesis, enlarged or not, may appear in a free or a bound informational shape. But – at certain stages of their development - not all languages dispose of the formal means to achieve the transition from free to bound. Let us once more look at the following examples:

Structure of meaning	**Possible English realization**
Action Synthesis: Men(Ag), rice(Pt), eat	
a) Conveying information unknown to the listener	Men eat rice
b) Not conveying new information to the listener	Men eat rice, you know. (They are usually healthy)
or: (are usually healthy	Men /who eat rice/
or: (are usually healthy)	/Men eating rice/

As pointed out by Hallpike,[10] primitive languages do not necessarily dispose of a ready-made formal shortcut to express the non-information synthesis. Instead they use devices like the above: 'Men eat rice, you know. (They are usually healthy)'. But even the devices used by highly developed languages like English, German, Chinese or Japanese are not to the same degree able to express all forms of the bound synthesis. English, for instance disposes of two different kinds of formal realization. It may repeat the substance formally realized as a noun by using a substitute usually called pronoun or it may express the bound synthesis without such a substitute:

Men /who eat rice/ are usually healthy   /Men eating rice/...

---

[10] http://www.gerojenner.com/wpe/the-hallpike-paper-universal-and-generative-grammar-a-trend-setting-idea-or-a-mental-straitjacket/

The Informational Structure of Meaning

This elegant abbreviation is possible in other cases too:

The wife /whom I married/ is now my best friend   /The wife I married/...
The man /to whom I gave the present/ is absent   /The man I gave the present **to**/...
The man /**from** whom I got the present/ is absent   /The man I got the present **from**/...

At some time in the historical development of English it was possible to omit the relative pronoun and to formally realize 'from' or 'to' at the end of the bound synthesis ('to' designates the receiver of the present, 'from' designates the agent ('He gave me the present').

But there is a price to pay for enhanced elegance. The abbreviated realization cannot express all kinds of synthesis. While it may be used in the first simple example below, it can no longer be applied in the three more complex ones.

a) The man /**to** whom I gave the gift/...
   the /man I gave the gift **to**/...

b) The man /**to** whom I gave the dog as a gift of gratitude/...
   the man I gave the dog as a gift of gratitude **to**/... *

c) The man /**to** whom I gave the gift I had received from Mary/...
   /The man I gave the gift I had received from Mary **to**/... *

d) The man /**from** whom I had got the present after dinner/...
   /The man I had got the present after dinner **from**/... *

The enlarged synthesis - or two bound syntheses as in b) - may no longer be realized without a substitute for the noun (the pronoun). This is true of the following instance as well:

e) The man, /whose dog has bitten a child/   no abbreviated alternative

Now, Chinese and Japanese are highly developed languages but they have no 'relative clauses', that is, no pronouns serving as substitutes. Formally realizing the bound synthesis not like English *after* the noun but *before* it, they can use the device without difficulty in a number of cases like:

The Informational Structure of Meaning

a) English: The man, /from whom I got the gift/, is my best friend

b) Chinese: /Wǒ dédào lǐwù de nánrén/ shì wǒ zuì hǎo de péngyǒu
　　　　　　/I　get　gift　　man　/ is　my　best　　friend

c) Japanese: /Watashi ni okurimono wo kureta otoko wa/ watashi no shin'yū desu
　　　　　　　/Me　to　　gift　　(Pt) gave　man　/　my　　friend　is

But more complex bound syntheses can no longer or only very clumsily be realized in this way:

a) English: The man, /whose dog has harmed my child/, is my best friend

b) Chinese: Zhège nánrén, tā de gǒu shānghàile wǒ de háizi, shì wǒ zuì hǎo de péngyǒu
　　　　　　That　man , his　dog harmed　my　child, is　my　best　　friend

c) Japanese: /Sono inu ga watashi no kodomo wo kizutsuketa otoko wa/ watashi no shin'yū desu.
　　　　　　/His　dog　　my　　child　(pt) harmed　man　/　my　　friend is

Chinese is forced to realize the bound synthesis as a parenthesis thus transforming it into a free synthesis. The Japanese rendering is very clumsy and therefore hardly used. The repetition of nouns by means of a pronoun used by Indo-Germanic languages is a formal device allowing much more freedom of expression than its abbreviated alternative though the latter makes for more elegance. English uses both modes, German does so to a certain degree but like Japanese and Chinese it places the abbreviated bound synthesis in a position *before* the noun it specifies:

English: /The enemies so much harrassing us/, could hardly be fended off

German: /Die uns so stark bedrängenden Feinde/, waren kaum abzuwehren
　　　　 /　us　so much harassing　　enemies/, could hardly be fended off.

The Informational Structure of Meaning

Practical definition of free versus bound synthesis

Form is no reliable indicator of whether or not a synthesis is free or bound. Practical clues based on semantic criteria should therefore be looked for, so that easy recognition is possible.

Generally speaking, the bound synthesis does not occur when the substance is *fully determined* nor when it is wholly *undetermined*. The speaker will use a fully determined substance like a proper name only in case he assumes it to be known to the listener - but then any further determination becomes superfluous. If it is nonetheless added, it must be new information and thus give rise to a free synthesis.

If, on the other hand, the substance is completely undetermined from the perspective of the speaker, and if he assumes it to be undetermined for the listener as well, any specification must automatically turn into new information and thus give rise to a free synthesis. Consider the following three examples:

a) Peter - who (by the way) arrived from London yesterday - will visit us soon.

b) Some person - who (by the way) arrived from London yesterday - will visit us soon.

c) The man, /who arrived from London yesterday/, will visit us soon.

Examples a) and b) are logically opposed in that the first refers to a fully determined person (proper name) and the second to a completely undetermined one. Nevertheless, both are identical in informational status. They contain a free synthesis realized according to the formal mode normally used for the realization of the bound synthesis (English relative clause). In both instances we may easily replace the English relative clause with the normal mode of realization for the free synthesis:

a) Peter arrived from London yesterday. He will visit us soon.
b) Some person arrived from London yesterday. He or she will visit us soon.

The third example differs from both its predecessors in that the speaker is well aware of the fact that among those men both he and the listener know there is one who arrived from London yesterday. So, the bound synthesis

/who arrived from London yesterday/ does not add new information to what the speaker and the listener already know. It merely serves to provide a reference. Just as proper names do not convey new information, in the same manner it is not contained in /the man who arrived from London yesterday/. *In fact, this expression is semantically on a par with a proper name.* It represents a device of language to name unique 'things' that are not, like proper names, expressed by means of special words.

However, proper names do not necessarily exclude the presence of a bound synthesis. If, for instance, the speaker assumes the listener to know (just like he knows for himself) several men by the name of 'Peter' he may well say:

the (or that) Peter, /whom you met in Brussels/...

In this - rather infrequent - case the formal sequence 'whom you met in Brussels' does not represent a free synthesis because 'Peter' is some definite person within a class of people bearing the same name.

No informational shifting within the bound synthesis

Normally, a bound synthesis represents in its entirety either the Topic or the Novum. There is no shifting within its boundaries.

/Small dogs/ like biting
/The hat on the table/ belongs to Paul
/The man walking along the road/ is a stranger
The man /who walks along the road/ is a stranger
Peter is a stranger

This statement seems to be contradicted by the following example, where the formant 'small' bears a particular stress.

Small dogs like biting

In fact, this is no longer a non-information (bound) synthesis. By *formal extension* (see IV,5) the same pattern is now used to express a conditional relationship. 'As long as dogs are small, they like biting'.

5 Semantic effacement

Any event spoken about by a speaker is semantically determined for him: It is either real or possible, belongs to the past or to the present, etc. But for the purpose of information it need not be so specified.

Take, for instance, the question: "Who goes home?" answered by the single expression: I. Here the word " I " presupposes the complete semantic structure "I, go, home". A synthesis (in the present case the abbreviated answer), *even if elliptical in form*, must be semantically complete since categories alone do not convey information. The listener knows that " I " does in fact means "I go home".

While *ellipsis* is a purely formal phenomenon, *semantic effacement* concerns specific parts in the structure of meaning. Effacement is very important, first, as a fact of semantics and, second, in its repercussion on formal realization. Let us consider the following complete bivalent action synthesis:

Semantic structure	**English realization**
Peter(Ag), Paul(Pt), a	Peter beats Paul

If we were to substitute 'somebody' (or 'something') for one of the substances, we express the fact of our reduced knowledge. We know for instance that somebody did beat Paul but not exactly who it was. (S_i = indefinite substance)

S_i(Ag), S(Pt), a	Somebody beats Paul
S(Ag), S_i(Pt), a	Peter beats somebody

Such a change from definite to indefinite substances still implies the assumption by the speaker that mentioning the agent (or patient) constitutes an important part of the information, even if one of the two is unknown to him. For this reason, these examples do not yet represent true instances of semantic effacement. The latter only occurs if the speaker takes it for granted that somebody must have performed the act of beating. He wants the listener to know about the mere event of Paul's having been beaten or that Peter applied beating to some person - whoever that may be.

a) ~~Somebody~~ beats Paul*   Paul was beaten   Paul got a beating

b) Peter beat ~~somebody~~*   Peter beat*

Semantic effacement of the patient is generally less usual than that of the agent. The reason for this lack of symmetry should be apparent. If we want to tell the listener something about an activity a certain person is engaged in, this information makes sense only if it is *definite*. We therefore say 'Peter beats <u>Paul</u>' (he did not beat Bill) or 'Peter cut a <u>tree</u>' (and not a log). If, however, we want to convey the information that Peter was engaged in some *unspecified* activity we say 'Peter <u>was working</u>' or 'Peter <u>was active</u>' and not 'Peter beat*' or 'Peter cut*'. In the two first examples we have effaced the specific patient, thus creating an unspecified action.

Semantic effacement generally removes a substance out of the focus of attention; it is therefore often, though not necessarily, coupled with *informational shifting*, that is, a change of Topic and Novum: the remaining substance may bear the focus of attention so that it assumes the role of informational novum.

In example (a) the effacement of the indefinite agent (~~somebody~~) has the result of making the patient 'Paul' the element that gets the focus and thus assumes the role of Novum. '<u>Paul</u> was beaten' answers the question 'who was beaten?' But such shifting does not necessarily occur. The same example when stressed alternatively 'Paul <u>was beaten</u>' answers the question '<u>what was done</u> to Paul?' So that the act 'was beaten' now represents the novum.

Languages differ in whether or not they allow formants expressing specific actions to be used in a nonspecific way by simply omitting the patient. In German we will easily say (I guess, more easily than in English) 'Peter <u>is riding</u>' (German: 'Peter <u>reitet</u>'). Unless we want to specify the patient (today he is riding the horse Bravado) it may remain semantically effaced because it is taken for granted. Both the speaker and the listener know that some horse must be involved in the act.

Effacement of the patient may occur if the latter is without interest within the context of information. Compare the following examples:

they stole (all the time)
they stole the garments

Only those parts of a synthesis may be semantically effaced that are *necessary* constituents of the real event and for this reason present in the minds of both the speaker and listener.

The Informational Structure of Meaning

But now consider the two following English examples:

a) They changed things rapidly

b) Things changed rapidly

The second sentence contains no reference to an agent; nevertheless, this cannot be viewed as an instance of semantic effacement since the real event need not have any human agent in the first place. Things may change by themselves. So, neither the speaker nor the listener must take such agent for granted.

In formal realization semantic effacement of the indefinite Agent may lead to quite different patterns. Japanese, for instance, does (or rather did) not use what is called a 'passive voice'; nevertheless, it has semantic effacement of the indefinite agent substance in some types of the subjective synthesis. (In the following examples the Patient of the subjective synthesis is rendered as psychic patient = P_{psy})

We(Ag)	see	the hill(Pt_{psy})		
~~Somebody(Ag)~~	sees	the hill(Pt_{psy})		
The hill(Pt_{psy})	can be seen		**Japanese**: yama(Pt_{psy}) ga	mieru
			The Hill	strikes the eye

I, he, we...(Ag)	find	a snake(Pt_{psy})		
~~Somebody(Ag)~~	finds	a snake(Pt_{psy})	**Japanese:**	
A snake(Pt_{psy})	was	found	Hebi(Pt_{psy}) ga	mitsukatta
			A snake	strikes the eye
			Hebi(Pt_{psy}) wo	mitsuketa
			A snake(Pt_{psy})	strikes (my, our..) eye

Note that both English 'the hill can be seen' as well as Japanese 'yama ga mieru' or 'hebi ga mitsukatta' are true cases of semantic effacement, and so is English 'A snake was found', but Japanese 'Hebi wo mitsuketa' usually represents a formal ellipsis since it presupposes some *definite* agent, the latter being made perfectly clear by the relevant context. So, the statement may be the answer to a question like 'I found a bird, what did you find?'

This is a good example for illustrating the difference between *formal ellipsis* and *semantic effacement*. Formal ellipsis is the omission in form of a semantic part made obvious by the context. In semantic effacement, the speaker withdraws his attention (and that of the listener) from any part of the synthesis that (being a *necessary* part of the event) is taken for granted by both.

We have seen that in a synthesis with two substances, the effacement of one of these may lead to *informational shifting* (change of topic and novum). Semantic effacement as such is, nevertheless, quite independent from informational shifting. This becomes obvious when a synthesis only contains a single substance, which is subsequently effaced.

Take the monovalent action synthesis. Semantic effacement of the agent is possible just like in the bivalent type:

~~people~~(Ag) were dancing (all over the place)

There was dancing all over the place

English has not developed a special type of formal realization for the monovalent synthesis with effaced agent, but German (and possibly other languages) have:

~~People(Ag)~~ were dancing

Es wurde getanzt

It was danced*

The element 'es' is a dummy formant in the sense that it does not convey any specific semantic meaning belonging to the logical or informational structure. But it does, of course, convey meaning as does any other part of any formal chain in natural languages. For this reason, it may not be omitted or replaced by any other randomly chosen formant like 'rem' or whatsoever. The *functional* meaning is indeed quite precise. In the above example, it indicates the semantic effacement of the agent, that is, the *absence* of a specific item of semantic meaning. Such dummy formants are the opposite of formal ellipsis. While the latter omits form where, logically speaking, a semantic content is *taken for granted* (because evident from textual or situational context), we now get a form but without a specific semantic content other than its abstract functional task.

Semantic effacement of the substance within a monovalent synthesis is not a necessary informational need of languages. It may be obviated by resorting to an indefinite substance instead of effacement. Thus, instead of saying 'es wurde getanzt' most languages will probably use an expression like '(some) people were dancing'. Obviously this alternative is of greater simplicity as it makes use of an established formal pattern for the monovalent synthesis. If, however, a special formal pattern has been created to permit the effacement in monovalent synthesis then this pattern can be extended (formal extension) to the bivalent type so as to create two formal alternatives for the realization of the latter.

a) Es wurde getanzt (It was danced*)

b) Indefinite persons(Ag)   ate   cakes(Pt)

c) *Es* wurde viel Kuchen(Pt) verspeist (*It* was much cake eaten*)
d) Viel Kuchen(Pt) wurde verspeist   (Much cake was eaten)

Type 1, the formal pattern created for semantic effacement in monovalent synthesis (es wurde getanzt) is in c) used for the bivalent type as well so that we get two alternatives of realization, a normal d) and a *derived* one c).

I mentioned before that there is hardly any informational use for effacement of *specific patients* ('he cuts*' instead of 'he cuts the tree', etc.). For the same reason the above German construction will hardly be used for effacing both substances though theoretically it could be put to such a use. There is just no (or at least much less) informational need to say something like 'es wurde geschlagen' (cutting occurred) instead of 'es wurde Holz geschlagen' (wood was cut).
English allows semantic effacement to occur with Action Syntheses where the action is enlarged by the use of an instrument:

a) The man(Ag) finally smashed the lock(Pt) with a hammer(Inst)

b) The hammer(Inst→*Ag*) finally smashed the lock(Pt)

The Informational Structure of Meaning

By semantic effacement resulting in formal ellipsis, the true agent disappears in the second example giving way to a pseudo-agent in the guise of an instrument ('Ist'). The latter is now felt to be somewhat active on its own exactly *because it occupies the formal slot* normally reserved for true semantic agents. This results in 'semantic tingeing' (see III,6), as formal realization makes the instrument appear like an agent.

Total semantic effacement and rank-lifting

Effacement may concern substances as well as their semantic roles. In all examples treated above the effacement of substances did not lead to that of the corresponding semantic roles.

a) Paul(Pt) was beaten   Paul(Pt) received a beating
b) Peter(Ag) reitet (Peter(Ag) is riding)

In (a) the substance representing the *agent* (for instance 'Peter') is semantically effaced but not the agent itself. In example (b) it is the substance representing the *patient* (for instance the horse 'Bravado') that is effaced but not the semantic role of patient as such. In both cases it would therefore be quite natural for the listener to ask for either semantic role: 'Who beat Paul'? or 'Which horse is Peter riding?'

Effacement may, however, be *total,* going beyond the substances and affecting their semantic roles as well. In this case, it is no longer natural to ask for either. Consider the following examples:

beating occurs frequently
love is universal
help is always useful

In general statements like these, nobody is likely to ask for either the agent or the recipient of help, love, etc. These statements are meant to be true regardless of any specific agents or recipients concerned. Any question as to the latter would therefore be rather unnatural. Effacement is here of a more radical kind. Not only the substances but also their semantic roles have been effaced. For this reason I speak of *total semantic effacement.*

The Informational Structure of Meaning

~~People(Ag)~~   beat   ~~people(Pt)~~   (frequently)

beating (is frequent)

The Topic as conceived by the speaker is reduced to the mere action of beating. Again, the *logical* structure of the synthesis remains unaffected. Both the speaker and the listener are well aware that persons in the role of agent and patient must be involved in the act of beating. It is precisely and only because both are *necessary* parts of the logical structure of meaning as a reflection of real events that agent and patient may be effaced. *Total effacement thus merely concerns the requirements of information.* The speaker wants to make clear that he has no use for any other content than that of the action to 'beat'.

Previously I have shown that effacement concerning substances may lead to 'informational shifting'; it now results in still another typical consequence. Total effacement reduces the synthesis to a mere action which, standing alone, *no longer conveys information*. The action as 'logical head' must therefore be followed by succeeding ranks ('beating <u>was frequent</u>' or 'beating <u>was very frequent</u>') or it must become part of a 'conjunction' (the beating <u>was cruel</u> = I, they, he etc. believe X. X = the beating was cruel).

The possibility of total effacement or using abstract concepts such as 'the beating', 'the hunt', 'the walk', 'love', 'danger', etc. is not realized in all language and does certainly not belong to its early genesis as it presupposes a specific formal device: the *lifting of semantic ranks into different formal slots*. It is therefore a mark of highly developed languages.

Total semantic effacement requires a specific mode of formal realization. In English and probably in most languages where it exists at all it is modelled on the basic type of formal realization but with a shifting of ranks.

I	I	II	III	IV	
Peter(Ag),	Paul(Pt),	beat,	frequently,	very	'Peter beats Paul very frequently'
I	II	III			
Beating,	frequent,	very			'Beating is quite frequent'
					(at this place)'

95

Rank II – semantically the action - is thus lifted into the formal position normally reserved for rank I, or the substance.

The lifting of rank II to the formal position of rank I entails further consequences: It leads to a *concomitant lifting of rank III to the formal position of rank II*. Thus 'frequently', which, in the first instance, belongs to the formal class of English adverbs ($_e$Adv), now becomes a member of English adjectives ($_e$Adj). And the reason why the formant 'very' cannot be satisfactorily classified on the level of form should be evident too: It is due to systematic rank-lifting.

Total effacement resulting in the lifting of ranks has an immediate and very important impact on the formation of word classes. Wherever it occurs, it may lead to a lifting of rank in formal realization and thus to totally changed paratactic word classes. The actual presence of a word class in English which unites substances (house, wood, person, thing, etc.) and actions (objective actions like beat, walk, cut; subjective actions like love, help, believe, etc.) in one and the same paratactic class (see V) is due to the fact that formal realization can no longer proceed in the same way as before since the positive formal element representing the agent does no longer occur. Therefore, the action itself is placed, or 'lifted', into this slot. The paratactic ordering of totally different semants (like substances. qualities, actions) within one and the same formal word class represents a progressive stage in the evolution of language.

Derivative use of total effacement

As a rule, total effacement is accompanied by zero-form for both agents and patients, that is these do not appear in formal realization after the action is lifted into the formal slot of rank I. But the reverse is not true: the lifting of rank, once introduced into a language, need not be accompanied by total effacement. In other words, the new device may be used to annul it.

Only in general statements of the type 'love is universal', 'help is useful', 'beating is frequent', the lifting of rank invariably tends to be accompanied by total effacement. But this need not be true for more specific statements:

	I	II	I	IV	III		
a)	Lumbermen (Ag)	cut	logs(Pt)	very	quickly		
			I			III	II
b) The lumbermen's(Ag)			cutting of logs(Pt)		is	very	quick

Here we observe the same phenomenon already described above with regard to German '*Es* wird getanzt' (~~People~~ are dancing). We have seen that the latter represents a new formal pattern created in order to realize effacement of the agent substance. Once established, such a new pattern may be used as a formal alternative to already existing ones even with no semantic effacement at all (*Es* wurde Kuchen(Pt) von vielen Leuten(Ag) gegessen). Such *formal extension* represents a very frequent linguistic phenomenon. A pattern once created is used for purposes other than those it originally fulfilled. In the present case, the normal pattern of English synthesis realization of the first example is duplicated by a formal alternative characterized by rank-lifting in the second instance *without semantic effacement*. Of course, in this new formal scheme, agent and patient can no longer occupy the same formal slots as before. New formal patterns must be established (agent and patient are denoted by a different case, the so-called genitive).

The lifting of rank, that is the transfer of semantic classes to a higher position in the formal chain, may exceed the boundaries of the single synthesis. Consider the following example:

Hatred is dangerous

On the semantic level, this utterance is composed of two syntheses, both subjective ones. 1) X = people hate people and 2) everybody fears X. Total effacement in the first of these leads to the 'subjective action' (to hate) being raised to a formal position corresponding to rank I, while the formal slot reserved for rank II is now filled with the term 'dangerous', which represents the second synthesis equally modified by total effacement.

It is not always an indefinite substance (somebody, something, people = S_i) that becomes effaced but quite often a *definite generalized* one (everybody, everything):

apperception synthesis:
Everybody holds X to be true   X is certain

affection synthesis:
Everybody fears X   X is dangerous

Remarks on form: the so-called passive voice in traditional grammar

A descriptive term cannot be deemed general, that is, of comparative use, if it only refers to a certain mode of formal realization to be found in some languages.
Japanese introduced a formal pattern which corresponds to our 'passive voice':

Many trees(Pt) have been fell-ed   Takusan no ki(Pt) ga bassai saremashita.
                                   Many   tree   fell - ed were*

but in former times the passive voice was hardly used, and up to now Japanese still makes extensive use of agent effacement linked to a type of formal realization that, *in form,* is entirely different from the 'passive voice'. Japanese uses an expression literally to be rendered somehow like 'the snake strikes the eye' where we would say 'the snake was found'

hebi   ga   mitsukatta
the snake   was found (fell in view *) or:

hebi   wo   mitsuketa
the snake   fell in my, your, his... view

Whether we take the Japanese or the English type of realization, the underlying semantic structure is characterized by the same process of agent effacement. The formal realization adopted in English is called 'passive voice', that of the Japanese example is not. For this reason, the traditional term 'passive voice' obscures the common semantic ground. Earlier our criticism against the term 'relative clause' was based on a similar objection. '(The) man going home...' and '(the man), who goes home...' are formal alternatives based on a common semantic structure, the bound synthesis, but the traditional term artificially creates two separate entities. For this reason, both terms are useless in comparative linguistics (General Grammar).

6 Semantic Tingeing

Semantic tingeing adds a difference of semantic nuance *resulting from formal classification*. It constitutes, so to speak, a feedback effect of form on meaning. The two following examples (one of which, though not ungrammatical, may be unidiomatic in English but is quite common in German) cannot be considered strictly identical in meaning.

| a) the tree | is high-er than | the wall |
| b) the tree | surpasses (German: überragt) | the wall |

In b) the relational quality 'high-er' is transformed into the relational semant 'surpass', realized in the same formal slot as actions. Although 'surpass' is identical in semantic content with 'higher than' the fact that it assumes the formal appearance of an English verb *induces* a *superimposed* semantic nuance, namely the more or less definite feeling that the tree in some way behaves in an active way like an agent. Semantic tingeing is one of the great means of poetic language. It allows playing with superimposed meaning. The same opposition is to be found in the following example:

the lion is stronger than the dog
in strength the lion surpasses the dog

Semantic tingeing is an all but universal fact of language. Quite a different example is provided by the prevalence of gender in most Indo-Germanic languages - English being an exception to this rule. It is an archaic remnant dating back to our prehistoric past when early philosophers were intent on classifying all things - not only living beings – according to the three qualities of male, female or neutral. Subconsciously we feel that there must be something feminine in 'la lune' and something more masculine in 'le soleil' (the opposite is true for German 'der Mond' versus 'die Sonne').

Gender distinctions have lost all practical value as concerns informational content. However, language ingeniously uses even traits without such importance. In poetic language the fact that a river is masculine in German (der Fluss) and a flower feminine (die Blüte) may play an important role because of semantic tingeing.

The Informational Structure of Meaning

7 Appendix: the Frozen Synthesis - the genesis of concepts

In all natural languages, the synthesis made of categories constitutes the basic building block. Now, in its bound state the synthesis not only behaves like a category but may actually 'be frozen' into a category.

The bound synthesis with an indefinite substance is equivalent to a more specifically defined substance:

a) Persons who drive a motor car are not admitted
b) motor-ists are not admitted.

German:
a) Personen, die ein Auto fahren, sind nicht zugelassen
b) Auto-fahrer sind nicht zugelassen.

English:
a) pieces that are broken (broken pieces) lie everywhere
b) fragments lie everywhere.
German:
a) Stücke, die gebrochen sind, liegen überall
b) Bruch-stücke liegen überall.

It seems quite likely that many specific concepts originated in this way but it depends on the prevailing pattern of formal realization, as adopted by a specific language, whether this origin may still be recognizable. While English prefers to realize its concepts using new non-composite formants ('fragments'), German is full of words built directly from the elements of a bound synthesis (like Auto-fahrer or Bruch-stücke).

Different types of synthesis thus become the catalysts, so to speak, in the genesis of concepts. Note that any synthesis (mostly connections, but even conjunctions) may become frozen and then behave like any simple category. We may therefore speak of 'frozen connections', too. Since this can be demonstrated particularly well in German, I will take most examples from my own mother tongue.

Quality synthesis
HOCH-SITZ (high-stand)
a stand, which is high, a high stand
KRUMM-ACHSE (crank)
an axis, which is crooked, a crooked axis

The Informational Structure of Meaning

Monovalent action synthesis
LÄU-FER (runn-er)
a man, who runs
FALL-BEIL (guillotine)
a chopper that falls down
enlarged synthesis
Schnell-zug (express)
a train, which runs quickly
Früh-aufsteh-er
a man who gets up early
with lifting of the action into rank I position
KINDER-GESCHREI (Children's crying)
after lifting of the action into rank I, 'children cry' becomes 'the crying of children'

Instrumental monovalent action synthesis
Wünschel-ruten-gäng-er (divining rod-goer)
a man who goes with a divining rod

Bivalent action synthesis
OPERN-SÄNG-ER (opera-singer)
somebody who sings operas
Hosen-träg-er (braces)
cords carrying trousers
DRUCK-KNOPF (push button)
a button which one pushes
enlarged action synthesis with semantic effacement of the patient
SCHNELL-KOCH-ER (quick cooker)
an instrument which cooks something quickly
with lifting of the action into rank I and semantic effacement of the agent
Fuchs-jagd
hunting for foxes
(after rank-lifting of the action and semantic effacement of the agent resulting in zero-realization 'people hunt foxes' becomes 'the hunting of foxes')
with lifting of the action into rank I and semantic effacement of the patient
Herren-jagd
(after rank-lifting of the action and semantic effacement of the patient; resulting in zero-realization) 'gentlemen hunt indefinite patient' becomes 'the hunting by gentlemen')

Instrumental bivalent action synthesis
Schlag-stock (bat)
a stick with which one hits somebody

Possession synthesis

The Informational Structure of Meaning

AKTIEN-BESITZ-ER (stock owner) (the possessor is focused substance)
man who owns stocks
MODERS-HOF   (the possessum is focused substance)
a farmstead owned by the Moders
with lifting of the action into rank I and semantic effacement of the possessor
Aktien-besitz (stocks owned)

Bivalent subjective synthesis
TIER-FREUND (animal lover)
somebody who loves animals

Conjunctions
logical conjunction
Freuden-schrei (cry of joy)
(after lifting the action as well as the quality into rank I and effacing the agent 'somebody cried because he was happy' becomes 'a cry because of happiness')
WARN-RUF (warning cry)
(after lifting the action to rank I and semantically effacing both agent and patient 'somebody shouts in order to warn people' becomes 'a cry in order to warn')

In a similar way Dakota allows different types of synthesis to be frozen. For instance, 'buffaloes that return running', 'tree standing on rocks', 'hill wearing blue robe', 'they who find a woman', etc. Perhaps the most frequently occurring type is the one derived from the localizing synthesis, which is chiefly represented by place names. Compare Kwakiutl 'island at the point', 'island in the middle', Eskimo 'the middle place', Tewa 'northern mesa where canyon is narrow' (cf. Boas in Hymes, 1964:174,175).

Any specific category like 'tree', 'divin-er', 'runn-er' may semantically come into being by acts of (intuitive) definition. But the same event may be defined in different ways. Whenever formal realization conserves the outward traces of such definitions, it is easily seen that one and the same 'real thing' may indeed be defined quite differently in different languages leading thus to different concepts.

Take for example English 'water divine-(e)r' and the corresponding German 'Wünschel-ruten-gäng-er'. In the first case the event is defined by the object looked for (water), in the second by the instrument used for finding it.

As far as meaning is concerned, new concepts mostly arise from definitions. This is quite obvious with regard to conscious processes. Thus, in science every new concept has to be introduced by way of definition. But it

The Informational Structure of Meaning

seems probable that basically the same process is involved in the unconscious genesis of concepts. If this is true, differences in formal realization between languages conserving definitions in outward form and others not doing so, cannot be explained merely with reference to semantics. Instead, they must be discussed within the context of the overall formal structure of definite languages.

The formation of composites may depend on the availability of certain formal means. Compare the remarks of Boas (in Hymes, 1964:174) on names of places in Kwakiutl and Eskimo. In Kwakiutl there are numerous terms for islands, which "refer to the location with regard to the neighboring land, such as 'island at the point', 'island in the middle', 'island in front', etc. In this language the term 'middle' is ordinarily a suffix and we find 'pond in the middle', 'hole in middle', etc., terms that cannot be formed in Eskimo except by long phrases that do not lend themselves well to the demands of a succinct nomenclature... Many of the locative suffixes of Kwakiutl are stem words in Eskimo, and since the nominal suffixes of Eskimo are attributive, the descriptive terms necessarily represent a different kind of imagery."

In other words, characteristics of formal organization in one part of the language (free synthesis) determine its shape in another part (frozen synthesis). The genesis of concepts, therefore, necessarily proceeds on diverging paths. Overall formal organization determines whether or not certain types of semantic definition may or may not find outward expression. This, surely, is a subject matter to be taken up by constructive linguistics.

The genesis of concepts is often linked to acts of definition regardless of whether or not the latter then becomes explicit in form. Formal explicitness (as exemplified by German) and formal implicitness (as in the case of English) both offer their own advantages as well as their particular drawbacks. Whereas English is terse and more elegant, German may sometimes become rather clumsy because of its rich vocabulary of composites. On the other hand, German offers a wealth of possibilities for the genesis of new concepts, possibilities not to be found in English. In German, the mere combinatory play with formants may, so to speak, induce new ideas. And it is, of course, much more difficult to express a new concept by creating an altogether new name than to use well-known elements.

Formal explicitness does not necessarily offer mnemotechnic advantages - the whole is mostly more than its parts - and sometimes it is quite different from the latter. The German composite 'auf-hören' (to finish) is made of the two formants 'auf' (up) and 'hören' (to hear). Somebody only knowing the meaning of both formants but not the meaning of the word 'aufhören' itself could be led to attribute a variety of possible different meanings to the composite. Most probably all his guesses would be far off the mark. As a general rule, it may be correct to say that in the majority of cases verbal composites in German cannot be intuitively derived from the meanings of their parts - consider such instances as 'auf-merken', 'durch-drehen', 'vor-sagen', 'ent-sagen', etc. So, these composites must be learned as *new meaning* just as the meaning corresponding to the English formant 'finish' has to be learned. Semantic derivation is, however, easier with nominal composites (Haut-creme = skin creme, Viel-weiberei = (many women) polygamie, Ent-völkerung = depopulation, Blei-vergiftung = plumbism, etc.)

From the comparative point of view, an important question arises as to the conditions that favor, or, on the contrary, tend to inhibit the formation of the frozen synthesis. Some languages use it extensively while others avoid it.

So far, I have not been able to go deeper into that matter. Judging by intuition (which, admittedly, can be quite misleading), I would expect languages relying on position in order to express semantic roles (for instance agent/patient) to be much more reluctant in the use of composites. The reason for my assumption is the following. Consider a sentence like English 'he drove a car experts long ago wanted to withdraw from circulation'. Here substances follow each other in immediate vicinity (car experts), their respective role being defined exclusively by position. A composite like 'car-expert' that creates the same pattern of immediate vicinity, is much more likely to come into conflict with this mode of realization than with the alternative mode used by languages making use of specific formants (article, suffixes or affixes) for expressing semantic roles (that is, languages like Russian, German, Sanskrit or Latin). Chinese a language relying on position even more than English corroborates this finding. Though it makes extensive use of composites these are of an altogether different type and meant to fulfil a different purpose (see VI,1).

IV The Formal (Symbolic) Realization of Meaning

Meaning is the foundation of language, which form is meant to 'realize', that is to transform in material signs susceptible of being exchanged between the members of a linguistic community. Meaning as such – i.e. mental images formed in the heads of speakers and listeners - is totally distinct from form, just as form – in the shape of acoustic waves or written letters - is totally distinct from meaning.

Any message based on the General Structure of Meaning present in the head of a speaker must be translated into Form, that is a structure of acoustic waves, in order to be received by a listener.

1 Formal means in natural language

These are, first, definite formal units like (a) mono- or polysyllabic sounds; second, their modification by (b) tones and (c) intonation and, third, (d) position that is different placements within the chain of units.

In natural languages semantic lexical items are for the most part *formally realized* by mono- or polysyllabic sounds called 'words'. Tones may, however, substantially reduce the number of sound units used for lexical items, as happens for instance in Vietnamese or Chinese.

Intonation is often used to distinguish the functional appearance of a synthesis as assertion or question, assertion or doubt. According to how I pronounce the German sentence 'Er kommt' (He is coming), it may be understood either as an assertion or a question. The same effect is produced by means of position, for instance 'Er kommt' (He is coming) versus 'Kommt er?' (Is he coming?) or finally by means of a specific sound particle as in Japanese: 'kuru' versus 'kuru ka?' the first meaning "(he) comes", the second 'does (he) come?'

The Formal Realization of Meaning

Position may be used as alternative formal device to express semantic relations like Agent and Patient (as in English: Peter hits Tom /Tom hits Peter) or Assertion and Question (as in German: Er kommt /Kommt er?) but it may not be used in natural language to express different lexical items in the manner of artificial languages.

2 Phonetics

describing the specific appearance of sound, tones and intonation will be left completely out of consideration in the present book as it does not concern the *relation* of the basic units of meaning (semants) to the basic units for form (formants) but represents a purely formal phenomenon below the level of formants.

3 The Differentiation-Value

We have now assembled the basic elements at the disposition of speakers when they express meaning by means of formal devices. But natural and artificial languages use form in quite different ways. Digital computer language expresses *all possible semantic differences* by different sequences (positions) of just two signs + and -. Natural languages, however, use (a) mono- or polysyllabic sounds, (b) tones, (c) intonation and, (d) position. In natural languages, these formal means can only express certain semantic categories as they have different "Differentiation Values".

For instance, what I call the '*Differentiation Value*' (Dif-Val) of *Position* is quite different in natural as compared to artificial languages. In the first this value is quite low while in the second it is almost infinite. The reverse holds true for *Sound Units* (words). In natural languages their Dif-Val may theoretically be limitless while it reaches its minimum in digital ones (+/-) because there are only two units. The Dif-Val of *Tones* seems to reach a maximum of six in natural languages. There are, for instance, five tones in Chinese (high: ma^1, rising ma^2, falling-rising: ma^3, falling: ma^4 and neutral: ma). Tones substantially reduce the number of elementary sound units (words appearing as syllables). Indeed, Chinese only uses a fraction of those needed by languages without tones.

The Formal Realization of Meaning

The Dif-Val of formal elements used in natural languages is responsible for the constraints operating in the formal realization of any possible language. In all of them basic elements of meaning (the semants listed in a lexicon) are formally realized by separate sound units, that is formants or words. Position rarely exceeds a Dif-Val of three. It may serve to formally distinguish statements from questions or Agents from Patients etc. But if position is used for the latter purpose as in English or Chinese it imparts a very special syntactic structure.

4 Formal Equivalence, Deficiency, Abundance

An interesting and intriguing aspect of natural languages is to be found in the fact that the formal realization of meaning may proceed in quite different ways. On the one hand, *identical* units of meaning may be formally realized in *different* ways, while, vice versa, identical formal means may embody more than one meaning. For instance, the above conjunction of two syntheses may be rendered in English in two alternative ways. 'Men eating rice are usually healthy' or 'Men, who eat rice, are usually healthy'. In Chinese only the first of these formal alternatives is admitted leading to a sequence like 'Eat rice men usually healthy'. The English case represents an instance of *different formal realizations of identical meaning*, in this case the bound synthesis.

The inverse case *of one and the same formal pattern expressing more than one meaning* is to be found in the so-called English 'relative clause'. The latter may express either an 'information' or a 'non-information synthesis'. Take for instance 'Peter, who (by the way) is a fantastic young lad, has my special approval'. Though identical in formal appearance to a bound synthesis, we are in fact faced with a parenthesis, that is an information- or free synthesis. The speaker wants to expressly inform the listener that he believes Peter to be a fantastic young man. He could have chosen the more usual formal realization: 'Peter is a fantastic young man. He has my special approval.

I will define the association of one semant (or group of such) with one formant (or group of such) as '*equivalent* mode of realization'. Similarly, realization will be '*abundant*' (nor redundant because it constitutes a linguistic asset) in the case of more than one formant for one semant, and it is '*deficient*'

The Formal Realization of Meaning

when there is less than one formant for one semant (= more than one semant for one formant). The distinction of synonyms and homonyms as deviations form equivalent realization is thus given a more general expression. And it does not only concern single semants but likewise groups of such as seen in the previous English example "who is a fantastic young lad" where one formal device (the so called relative clause) demonstrates deficient formal realization as it may be used to express two different contents of meaning (a free or a bound synthesis). 'Flying planes can be dangerous', one of Chomsky's example provides a further case for *formal deficiency*. The two syntheses, one a Psychic-state Synthesis: It is dangerous, we find it dangerous etc. and an Action-Synthesis:

Structure of meaning (conjunction)	Possible realization in form
There are flying planes They can be dangerous	Flying planes can be dangerous
If (when) somebody flies a plane That can be dangerous	Flying planes can be dangerous

In both examples the Agent of the Action Synthesis "fly, planes" is suppressed. It does not matter who directs the plane but the Agent flying the plane is different from ourselves who perceive these flying planes as being dangerous to ourselves. In the second instance the Agent is suppressed as well. It does not matter whether we or somebody else flies the plane but it is understood that the Agent who flies the plane is identical to the one who perceives the matter as being dangerous. It is, of course, the context that determines the right meaning

These are exotic examples of formal deficiency. More usual are cases like board (= plank) or board of directors where again the context provides the clue for meaning.

Formal abundance occurs much more often though minor shades of meaning usually tend to differentiate even between apparent synonyms (peasant, farmer; since, because etc.).

But there are many instances where no semantic differentiation occurs.

had see-*n*, *had* walk-*ed*, *had* cut-*0*;

Here we are faced with two different kinds of formal abundance. First, abundance is expressed by the fact that the semant in question (a certain temporal specification which we need not analyze at this point) is discontinuously realized by means of two separate formants: 'had ... -n' or 'had ... -ed'. This is an instance of synchronous abundance since both formants must be present in order to formally realized one semant. But these examples present a second type of formal abundance too, a contextual one, because the right hand formant may appear in different shapes as '-n', '-ed' or '-0' according to context.

The formal realization of the plural in modern English provides a further case of abundant realization (which in the distant past may have quite different semantic implications):

house-s
ox-en
child-ren
mice

The last instance (mice) simultaneously represents a case of 'formal deficiency' (more than one semant being realized by just one indivisible formant: mice).

Even entire syntheses may appear at the same time as instances of formal abundance or formal deficiency. The so-called relative clause in English exemplifies formal abundance as 'Men eating rice... ' and 'Men who eat rice' both express the Bound Synthesis. But it is a case of formal deficiency too because in sentences like Peter, who is a formidable young lad... ' the same pattern represents a Free Synthesis.

Needless to say that equivalent realization is the general rule in all natural languages.

5 Formal Extension

This is a device to be found in every language and in most it seems to be used to a great extent. Formal extension is a more general term for both formal deficiency (one formant or group of such used for more than one semant or group of such) as well as for formal abundance (more than one formant or group of such used for one semant or group of semants).

Formal extension through formal deficiency
This device is found wherever one formant or one formal pattern is applied to cover semantically different instances:

1	2	3
a) he	hits	(the) foe
b) Paul	likes	(his) brother
c) (the) tree	tops	(the) house

In all three cases the first position is in traditional grammar named Agent, the third is the Patient. If General Grammar only admits the first example as a real instance of an Agent exercising a physical influence on the Patient, then b) and c) cannot be designated in the same way. In b) we have a Psychic State Synthesis with a psychic Agent and Patient (Ag$_{psy}$ and Pt$_{psy}$) and the third example represents a spatial relationship. Other languages than English do indeed differentiate between these semantically diverse examples. Instead English formally extends the pattern of a) to b) and c) - and to many more semantically different relations. Formal deficiency thus turns into *formal economy*. Instead of multiplying formal schemes wherever a semantic relation deviates from an original pattern, this pattern is maintained in order to simplify language.

We will see later (cf. VI,1) that formal equivalence must be a law in the realization of semants by formants. Here formal deficiency (using one formant for more than one semant) can only be an exception because it would *contradict formal economy*. But formal deficiency contributes to formal economy when applied to formal patterns as the above shown.

Formal extension through formal abundance
English like other Indo-Germanic Languages realizes tense and number by way of formal abundance, here turned into formal redundancy, instead of formal equivalence

a) *Yesterday* he climb-*ed* the mountain   formal abundance (redundancy)
b) Yesterday he climb the mountain*   formal equivalence
c) *He* climb-*s* the mountain   formal abundance (redundancy)

The Formal Realization of Meaning

d) He climb the mountain* formal equivalence

French and German are still more redundant (namely in gender, tense and number). The corresponding Chinese sentence has no obligatory rendering of gender, tense and number: The synthesis is realized without redundancy in formal equivalence (semantic structure: Pural girl, come /girl, bad/).

The bad girl-*s are* coming	number: twofold formal abundance
Die böse-*n* Mädchen komm-*en*	number: threefold formal abundance
Les mauvai-*se-s* fille-*s* vien-*nent*	gender: twofold formal abundance
Plu bad-fem-plu girl-plu come-plu	number: fourfold abundance in written, twofold in spoken language
Huài nǚhái men lái-le	formal equivalence
Bad girl plu come	

Formal abundance does not generally lead to redundancy, often it contributes to the wealth of language as was the case when the Normans added many French words to the Anglo-Saxon stock and thus enriched the existing idiom. But this is true as well when one formal pattern is used to express more than one semantic pattern as was shown in the English example, where the relative clause represents a free synthesis:

1) Mr. Abbot, who by the way is on the way to London, is my personal friend

Here the pattern normally used to realize the bound synthesis (relative clause) is in fact made to express a free one as a parenthesis. The language is made richer by providing more than one formal alternative. The same can be said of the German example already mentioned earlier as a means to suppress both the indefinite Agent and Patient. Here a formula originally and indeed mostly used to efface both Agent and Patient can be used to reintroduce both.

a) Indefinite persons(Ag) dined indefinite(Pt)
b) *Es* wurde gespeist (It was dined* = ~~People~~ were dining)
Formally extended to:
c) *Es* wurde viel Kuchen(Pt) verspeist (*It* was much cake eaten*)
or even to:
d) *Es* wurde von den Gästen(Ag) viel Kuchen(Pt) verspeist (*It* was much cake eaten by the guests*)

6 Morphology

Semants may be formally realized as independent or dependent formants. Thus, in English the Action Synthesis "We, run" is expressed by means of two independent formants as 'we run', while Italian fuse both semants into a single formal entity 'corr-iamo' where no part may occur independently. In other cases, an independently occurring formant like 'cloud', 'tree' etc. may be combined with a dependent one denoting plural: "cloud-s', tree-s' etc. In some languages such combinations appear as suffixes, in others as prefixes. Dependent formants may be attached to Substances like tree, cloud etc., to Qualities like French bon or bonne according to gender and to Actions specifying temporal or other characteristics (English 'walk' versus 'walk-ed'. These may be merely formal choices bereft of any added semantic content. For instance, there is no change whatsoever in semantic content whether we say English 'We run' or Italian 'Corr-iamo'. Likewise, no difference in semantic content is implied when the Action Synthesis "We, climb, past" is expressed in a formally equivalent way (one formant for one semant: we climb-ed) or in a formally deficient manner (less than two formants for two semants) like in English "We, run, past", which is realized as 'We ran'.

In natural languages, only a narrow range of semants are eligible for the role of dependent formants (affixes). Open-field semants like Substances (house, stove, deer, tree etc.) never appear in this role nor do open-field Actions (like run, climb, read, toil etc.). The same holds true for open-field Qualities (like hoarse, blue, tough etc.). These classes comprise a theoretically *infinite* number of members, which do not semantically modify each other. Formal realization as dependent formants is strictly limited to *closed-field semants* like:

Temporal (present/past etc.)

Spatial relations (here/there etc. like in some Amerindian languages)

Number-related (singular/plural often extended to Qualities)

Person-related (I/you/ etc.)

Agent-patient related (Latin equ-us/ equ-um, Amerindian languages call-he-him etc.)

Comparison-related (fast, fast-er, fast-est)

The Formal Realization of Meaning

Category-related (teach, teach-er = Action changed into substance; must not be confounded with rank-lifting: good, good-ness, see II,5)

Social status-related (like in Japanese addressing equal, higher or lower persons etc.)

Gender-related (masculine/feminine etc.) and similar polar (closed field) semantic categories.

In most cases the use of dependent formants does not as such imply any semantic differences. In so far as this is true, we are faced with purely formal alternative. It then does not matter whether a language is analytic, that is predominantly consisting of independent words or synthetic (agglutinating, oligo- or polysynthetic). But in some cases, such difference does indeed matter. For instance, most Indo-Germanic languages distinguish the gender of substances and even extend this distinction to Qualities etc. Undoubtedly this differentiation has a remote origin in a philosophy that stressed the difference of sex (from which it is evidently derived). So, in the beginning it did express what we may call a definite world view. Edward Sapir and his followers are right when they try to derive different attitudes vis-a-vis reality from the formal appearance of language. Or rather they may be right in a limited number of instances. As soon as the original distinction of sex turned into gender so that now German spoon had to be masculine (der Löffel) why German fork (Die Gabel) had to be feminine and German knife became neuter (Das Messer), philosophy got lost in the way of mindless systematization. Poets may still play with the different genders of moon and sun but in average huge the distinction of gender is nothing more than an addition and superfluous lexical item without any semantic significance (it has been all but abolished in English).

Morphology may, however, play a very important role in formal realization - it does so, for instance, in English. Under the heading of *'Differential Analysis'*, I want to devote some lines to a very specific case of English morphology – indeed, the very core of English grammar -, which up to now has been completely overlooked (arguably because of its complexity).

The Formal Realization of Meaning

Differential Analysis

Descriptive linguistics on the syntactic level would be much simpler if the relationship between meaning and form strictly conformed to only one type: that of formal equivalence, i.e. one formant for one semant. But often one semant is formally expressed by more than one formant (formal abundance) or less than one formant (formal deficiency). The empirical study of languages shows that formal deficiency often occurs right in the syntactic core. Lacking the appropriate general terms, traditional grammar was all but incapable of describing such intricacies. *Differential analysis* is a method explaining the formal organization of meaning where it is not intuitively evident, that is, in cases of non-equivalent realization. In other words:

Differential analysis is an exact method for mapping the semantic deep structure of any language onto its formal surface. Insofar as the chosen deep structure is common to several or all languages, it represents the only scientifically sound method of comparative linguistics.

It is certainly the most precise instrument of analytic linguistics developed until now. Differential analysis proceeds in four clearly defined steps.

First step, determine the semantic structure. In chapter *General pattern of formally realizing the bound synthesis* the semantic deep structure consisted of three basic distinctions: a) agent/ patient, b) free/ bound synthesis, c) agent in the role of topic/ patient in the role of topic. If all three oppositions are to be formally expressed, the minimum number of formal examples will be 2 x 2 x 2 = 8. But in English the actual number is at least twice as high because of purely formal alternatives (man beating Peter = man who beats Peter) etc.

Second step: take sentences that differ *by just one* of their semantic contents (semants) but are identical as to all others.

Third step: replace the formants that realize this difference with that semant (semantic content) and repeat this for all semants (in this specific case for all three dichtomies).

Fourth step: for each formant show all semants arrived at in step three. You will then discover that one formant may simultaneously or alternatively realize more than one semant. Some of these semants will be active in one context and suppressed in another, some will be realized synchronously with others, and some not. In this way we get an exact knowledge of which for-

The Formal Realization of Meaning

mants realize which semants. In other words, we have uncovered the hidden deep structure of meaning as it appears on the surface, and we understand the complex workings of the human brain. *The ultimate aim of replacing arbitrary formants with non-arbitrary semants is to allow intra- and interlingual comparison.*

Differential analysis is no mere addition to the procedures of traditional grammar - it revolutionizes its very foundation, since it makes sense only if based on pure meaning - the rock on which all formal realization is built. From a logical point of view, it would be circular to ask how formal or hybrid terms (i.e., terms defined partly by form and partly semantically) map onto the formal surface of a given language. In chapter *General pattern of formally realizing the bound synthesis,* this was shown for hybrid traditional terms like S, O, main/relative clause, passive, etc. These terms had to be replaced by purely semantic ones.

The following represents a rigorous application of differential analysis. Consider the following English sentence:

(the) door is be-ing open-ed

At first glance, we are able to distinguish the following semantic contents, which we will have to abbreviate as follows:

S(Pt) figures as a substance (door) in the role of patient,
b (open) represents the bivalent action,
pr / npr is meant to indicate progressive / non-progressive,
c / nc refers to completion versus non-completion of the action
ta / tb / tc present versus past and future time.
s / pl = singular versus plural
ac / pa expresses semantic inversion (English 'Active' / 'Passive voice'),

However, we can by no means be sure, which formants exactly realize which semants. Look, for example, at the formants 'is', 'be', '-ing' and '-ed'. The task of differential analysis is to assign its proper meaning to each formant within the chain. Obviously, the seven above mentioned semants to not suffice for our task. Comparing 'the door is being opened' with 'The door being opened (was the second to the right), we realize that the formant 'is' not only realizes

The Formal Realization of Meaning

singular number and present tense but stands for the free synthesis as well. To the seven semantic elements, the semants, belonging to the *logical* structure of meaning we must therefore add a *functional* meaning: that of the free synthesis as opposed to the bound one, which for more convenience will here be distinguished by the symbols '!' and '//' respectively. So, the above chain must be described by means of a total of eight semants, that is

S(Pt), b;   pr/npr, c/nc, ta/tb/tc, s/pl, ac/pa;   ! or //

These must, in some way, realize the English formal chain:

(the) **door** is be-ing **open**-ed

and its semantic variants. However, apart from two out of these six formants, namely 'door' and 'open', representing S and b respectively, it is quite impossible to attribute any definite semants to the four remaining ones, that is, to 'is', 'be', 'ing' and 'ed' in an intuitive way - this would be nothing more than mere guesswork.

It is differential analysis which allows us to surmount this difficulty *by comparing all instance with just one semant having a different value.* We thus compare 'the door is be-*ing* opened' with 'the door is opened', which reveals the semantic element of progressive versus non-progressive. Comparing 'the door *is* opened' with 'the door *was* opened' we get the semantic element of time. By contrasting 'the door *is* being opened' with 'the door opened (just now does no longer serve any purpose)' we find that the formant 'is' expresses the free synthesis in the first instance. By comparing 'the door is being opened' with the doors are being opened' we get the semantic element of number. When contrasting the two sentences 'he opened the door' and 'he has opened the door', we conclude that in the second case a certain result has been achieved and still persists, namely that the door is now open. In the first instance we do not imply that the result still persists at present. The door may have been shut meanwhile. Repeating such differential comparison, we finally extract all logical and functional semants in question within the *entire formal chain* over all its possible variations.

As it turns out, formal organization in natural language is much more complex than we would expect. Indeed, the solution to our problem takes the following shape:

The Formal Realization of Meaning

$$\text{engl} \longrightarrow \begin{array}{llll} \text{door} & \text{is} & \text{be-ing} & \text{open-ed} \\ \textbf{S(Pt)} & \text{pa}_1\text{:ta:s:!} & \text{pr}_1 - \text{pr}_2 & \textbf{b} - \text{pa}_2 \end{array}$$

It can be seen from this semantic analysis that 'nc' (non-completion) is realized by zero-form, that is, by the absence of any positively realized formant. Semantic inversion (English passive voice) is expressed simultaneously by two formants pa_1 and pa_2, that is, by *formal abundance*, while the single formant 'is' realizes four semants at the same time (*formal deficiency*).

Even so, the picture of formal realization resulting from the above non-contingent expression is still too simple. As soon as we add those semants which, in other contexts, may be expressed by the formants in question but *are suppressed in the present environment*, we arrive at the complete formula b) below (with suppressed formants put into brackets):

$$\begin{array}{llllll} & \text{door} & \text{is} & \text{be} & - \text{ing} & \text{open} - \text{ed} \\ a)_{\text{engl}} \longrightarrow \textbf{S(Pt)} & \text{pa}_1\text{:ta:s:!} & \text{pr}_1 & - \text{pr}_2 & \textbf{b} - \text{pa}_2 \\ b)_{\text{engl}} \longrightarrow \textbf{S(Pt)} & (\text{pr}_1)\text{:pa}_1\text{:ta:s:!} & \text{pr}_1 & - (//)\text{:pr}_2 & \textbf{b} - (\text{tb})\text{:}(c_2)\text{:pa}_2 \end{array}$$

Analyzing the formant 'is' in 'he is open-ing the door', we find that together with the formant 'ing' it expresses the progressive. This semant is, however, contextually suppressed in our main example and therefore set in brackets (pr_1). Comparing with 'play-ing children (must be protected)', where the formant 'ing' expresses the bound synthesis, we see that this functional semant too is suppressed in our example and must be put within brackets (//). In other cases, the English formant '-ed' contributes to the expression of completion or it indicates the past, both semants are suppressed in our example: (c_2):(tb).

Such mathematical analysis of language may seem to most readers rather superfluous if not outright repulsive. But my and your brain, dear reader, must perform such an analysis otherwise we would be unable to understand the basics of language. Thus, the preceding demonstration only makes explicit what the brain achieves in a totally unconscious way. No native speaker of English has the least difficulty in properly assigning the right

The Formal Realization of Meaning

semants to formants like 'is', 'ing', '-ed' despite the fact that there meaning changes substantially according to the contextual situation.

V Syntax and Paratax - Basic Modes of Formal Realization

Syntax and Paratax are logical counterparts since one cannot exist without the other as seen in the following examples:

	Class 1	Class 2	Class 3	Class 4
	Paul	strikes	the	tree
(Her)	Screaming	hits	my	ear
	Brightness	dazzles		us
	Happiness	transports	the	lovers
	Fortune	expulses		envy
	Swiftness	decides	our	victory

As long as we turn out attention merely to the succession of words in a single sentence, we deal with Syntax. As soon as we ask about possible substitutes for each item, we view these as a class and thus highlight Paratax.

In the preceding scheme, the two formal classes of nouns (1 and 4) apart from containing substances as in the first example comprise Actions, Qualities and Psychic States.

Not only is a purely formal definition of nouns impossible, the same applies to other categories like Agent and Patient if we consider their actual semantic content in a given language. Just consider the following three English examples:

a) Peter(Ag) beats Jim(Pt)

b) He(A_{psy}) likes the girl(P_{psy})

c) The house tops (is bigger than) the tree

A true Agent where one substance acts on a second is found in the first instance only. The second example describes an inner psychic state of one person with regard to another, while the third illustrates a spatial relationship between two substances. While there is a simple form for the first instance in all languages, only a few extend the basic formal realization for true Agent-Patient relations beyond its proper field to include examples like b) and c). This means that in some given language like English Agent and Patient do no longer constitute purely semantic categories but are partly defined in a formal way. In the last two cases there exists an overlaid semantic tingeing as if the first noun played something like an *active role* with regard to the second (in other words: the basic formula developed for a) exerts a hidden influence on derived examples like b) and c)). For overlaid or imposed *semantic tingeing,* see II,6.

At this place, I want to stress that entirely different semantic classes like Substances, Actions, Qualities, Psychic States etc. may all be realized as members of the same formal class - in the present instance as members of English nouns ($_e$Nouns). Paratactic classification is, however, quite different in a language like Chinese. This holds true for $_e$Agents and $_c$Patients as well If, nevertheless, traditional grammar uses the same word 'noun' in English as well as in Chinese, it is because English and Chinese nouns partially overlap with regard to their semantic contents: both, the Chinese noun ($_c$Noun) and its English counterpart ($_e$Noun) contain substances as their main item. And this applies to English and Chinese Agents and Patients and to all terms of traditional grammar defined both through meaning and form.

So, let us always think of Paratax and Syntax at the same time as the successive elements of the latter are the members of paratactic classes. Both are part and parcel of formal realization – or rather they constitute its very basis. And both are *specific* for each language. It has been explained that differences in paratactic order are largely caused by 'rank-lifting' (see III,5).

1 Shortcomings of Chomsky's Generative Grammar

Since paratactic formal classes are language-specific, there can be *no nouns as such* but only English, Chinese, Japanese nouns, adjectives, verbs etc. (each formal class filled with different semantic classes). English but not Chinese nouns comprise semantic members like giftedness, extraordinariness etc. This means that the English Noun must be distinguished from its Chinese counterpart by an appropriate notation, for instance $_e$Noun versus

$_c$Noun - and so on for all remaining formal categories. This is an item of utmost importance, when it comes to evaluating the possible performance of Chomsky's Generative Grammar. *Only if the semantic contents grouped in paratactic classes like verbs, nouns etc. were identical in all human languages*, would Generative Grammar as conceived by Chomsky make any sense. If not, Chomsky's Generative Grammar, instead of explaining the variety of languages, actually *explains it away*.

Generally speaking, Paratax as a basic and distinct procedure in every language has been all but overlooked in Traditional Grammar with the exception of 'Distributional Analysis', which, however, discarded meaning so that it cannot produce any results in the field of comparative linguistics.

2 Usefulness of traditional terminology

We could imagine an *ideal language* – never found in actual use – where the formal class Noun would only comprise the semantic class of substances; Verbs would only contain actions; Adjectives only qualities; Subjects would only represent agents and Objects would exclusively refer to patients and so on. On the basis of such an assumption, we would be justified in using the traditional notation.

Real languages tend to be infinitely more complex. In the present work, I will refer to formal Paratax only in view of refuting Chomsky's assertion that Nouns, Verbs, Nominal Phrases etc. may serve as general categories in the description of language. In other words, the existence of formal Paratax proves the need for linguistics to base generativeness on the more deep-lying categories of pure meaning and its possible formal realization.

When turning to the analysis of some given language, any *comprehensive description* of formal Paratax would, however, be quite useless as any native speaker has an intuitive grasp of the matter. Instead we should ask and answer a different question. *By what mechanism* do certain languages like English succeed in putting quite different semantic categories like Substances, Qualities, Actions etc. in the same formal paratactic class of English Nouns as described in the above mentioned example? I have discussed this problem under the head of rank-lifting (see III,5).

Syntax and Paratax

VI Law in Language

The primary task of a linguistic science that differs from its intuitive understanding and the natural joy that its use conveys to the receptive mind, is to establish the limits of law and arbitrariness. The laws governing the human mind when it dissects reality and creates the Types of Synthesis in the Logical Structure of Meaning belong to Psychology. They are here accepted as facts. The Devices of Information in the Informational Structure of Meaning are the proper field of Linguistics. This equally applies to the laws governing Formal Realization.

Are there any laws at all - laws that in the bio- and psychological realm we should perhaps rather designate as regularities? The Swiss linguist Fernand de Saussure distinguished between signifiant and signifié that is, in the present terminology, between formants and semants. He found out that in relation to the second the first are purely arbitrary. The semant "tree" for instance may in different languages be formally realized as 'ki' in Japanese, 'tree' in English, 'Baum' in German, 'mu' in Chinese, 'arbre' in French and so on. In other words, it may take any acoustic shape whatsoever. The relationship of semants to formants is definitely arbitrary, no regularity can be shown except for the rare cases that languages make use of so called onomatopoeia. The cry of cocks thus becomes 'Kikeriki' in German, 'Cocorico' in French, Cock-a-doodle-doo in English. But these are exceptions. De Saussure's statement holds true for the overwhelming majority of cases.

But does it hold true as well for the relation of meaning to form beyond the level of semants and formants? By no means. Let us make our point quite clear. Language is to a large part the product of (collective) mental spontaneity being in many regards marked by contingent history and pure chance. But it contains certain features, which we are able to prove as lawful because subject to definite constraints.

Law in Language

1 First basic law (concerning formal equivalence)

There is no a priori reason why one element of meaning (semant) should be expressed by just one element of form (formal equivalence).

Consider the two following alternative modes of realization, the first being equivalent, the second formally deficient (one formant for more than one semant):

Semantic Structure	Formal realization1 equivalent	Formal realization 2 deficient
"Man, approach"	'(The) man approache(s)'	'Deng'
"Man, recede"	'(The) man recede(s)'	'Ding'
"Man, sleep"	'(The) man sleep(s)'	'Dang'
"Man, read"	'(The) man read(s)'	'Dung'
Total number of semants = 5	Total number of formants = 5	Total number of formants = 4

The formants 'Deng', 'Ding' etc. are, of course, arbitrarily chosen. They may be replaced by certain cries of animals when they feel threatened or by specific gestures in sign languages. The examples prove that equivalent realization is less economical (5 formants) than a deficient one (4 formants). However, the break-even point is already surpassed if we *just double the number of messages* adding the following instances:

"Woman, approach"	'(The) woman approache(s)'	'Däng'
"Woman, recede"	'(The) woman recede(s)'	'Düng'
"Woman, sleep"	'(The) woman sleep(s)'	'Döng'
"Woman, read"	'(The) woman read(s)'	'Dyng'
Total number of semants = 6	Total number of formants = 6	Total number of formants = 8

Here the number of formants needed in deficient realization already surpasses that for its equivalent counterpart. This explains why only primitive animal or sign languages transmitting hardly more than a handful of messages belong to the deficiently realized type. Semantic structures like "Come to our village!" or "Flee beyond the mountain!" may then be expressed by some single indivisible sign.

The advantage of deficient formal realization is, thus, soon offset when the number of semants increases. In fact, the number of formants required in formally deficient realization increases exponentially with a growing number of semants.

So we may establish our first law of formal realization:
Every developed natural language, that is every language equipped even with a very modest vocabulary, is bound to realize single semants by means of single formants, that is according to formal equivalence.

Human memory would be totally overloaded if this rule did not apply. But this does by no means exclude deviations from optimal economy of formal realization both through deficiency or abundance. The normal pattern for realizing the past in English is provided by the suffix '-ed' (he lik-ed, explain-ed, jump-ed) but in a number of cases formally deficient realization occurs (he sang, came, brought, left). Cases like 'board' used for plank and in 'board of directors' illustrate a further type of formal deficiency. Other examples are stalk, left, skate etc.

Formal abundance (more than one formant for a single semant) likewise constitutes a deviation from optimal economy of formal means. As a rule, the English plural is realized by means of the suffix '-s' (house-s, train-s, cloud-s) but in some cases it is replaced by '-ren' (breth-ren, child-ren). One semant is thus realized by more than one formant. English comparative is realized by the suffix '-er' (bigg-er, hard-er) and by the preposition 'more' (more extensive).

Formal deficiency may be synchronous (when two or more semants are *simultaneously* evoked by a single formant), or it may be *non-synchronous*. In this case it depends on the context which of two or more possible meanings is expressed. When it is neither the one nor the other, it is neutral.

The English formant 'they' always expresses the third person together with plurality, that is, in a synchronously deficient way. Japanese expresses the same meaning by using two formants (kare-ra), one for the third person (kare) and one for plurality (-ra). It therefore adopts an equivalent mode of realization. English 'stalk' provides an instance of non-synchronous deficiency: It depends on the context which semant is understood.

German provides an example of synchronous abundant realization. While English realizes the meaning "finish" in a mode of formal equivalence: one formant for one semant (he *finished* the work long ago), German here adopts a mode of synchronous formal abundance (er *hörte* mit der Arbeit vor langem *auf*): the two formants 'hören' und 'auf' are used simultaneously in order to express one semant.[11]

We have already seen that the two possible modes of deviation from formal equivalence are not restricted to single semants but can be found in syntheses and conjunctions (combined syntheses) as well. The Chomskyan example 'Flying planes can be dangerous' is an example of formal deficiency covering a conjunction. *According to the context*, the expression 'flying planes' may be understood either as: 'It may be dangerous to fly planes' or as 'Planes that fly may be dangerous'. Formal abundance beyond the level of words is to be found in the two alternatives 'the man going home is my friend' and 'the man who goes home is my friend'. This is an instance of neutral formal abundance.

Deviations created by formal deficiency or abundance may amount to several dozen perhaps even some hundred - nevertheless, they are without significance compared to those astronomical numbers if overall formal deficiency would be applied in the way illustrated by the first above-mentioned examples. *In other words, the law of formal equivalence only allows for minor exceptions.*

2 Second basic law (formal realization of open field semantic categories)

Open-field semantic categories like substances, actions, qualities may only be realized in form by open-field formants. Thus, substances like tree, cloud, stone, man, dog … may not be realized by position as happens in artificial binary systems. In any natural language they must be realized by so many different formants usually called words.

---

[11] Etymologists will probably be able to prove that in some more or less distant past the two formants combined in German 'auf-hören' evoked independent meaning. At the time of its formation the expression must have had a composite meaning corresponding to equivalence of realisation. But nowadays -- and synchronous analysis deals with the actual state of a language - 'aufhören' has no other meaning than English 'to finish' (even if, to however small a degree, the semantic contents of its two components should still linger as a kind of <semantic tingeing> in acute linguistic awareness).

At first glance, Chinese seems to be an exception to this general rule because it uses a *limited* number of formants – only about four hundred – together with a quite restricted number of tones, namely four. For this reason, the total number of formants distinguished by different tones cannot exceed 1600 – which, obviously, does not suffice to realize a potentially infinite number of semants. In order to overcome this limitation Chinese resorts to an extensive use of *synchronic formal abundance*. It expresses one single semant by means of two formants that are often quite similar in semantic content. Eye is, for instance, expressed as yan-jing, where both yan and jing separately refer to the eye. By this simple device, the Chinese language may enlarge the number of possible formants beyond all limits of practical use to 1600 times 1600.

3 Third basic law (the use of position in natural languages)

If used at all as a means of formal realization, *position can only apply to closed-field semantic categories* like Agent versus Patient, Question versus Statement and similar polar semantic categories. In English the semantic difference of the two sentences 'Bill hits Paul' and 'Paul hits Bill' is due exclusively to position, as is true of German 'Kommt er?' (does he come?) when compared to 'Er kommt' (he is coming). Comparative linguistics shows that position is used only exceptionally to realize closed-field semantic contents, but, when being so used, it plays a prominent part - for instance in English as well as in Chinese where it is used to contrapose Agents and Patients. As this opposition pervades the whole language, it gives rise to the distinctive formal structure of both languages.

Position in natural language is, however, only used with a Differential-Value of two, it doesn't even extend to three positions as testified by English and Chinese, the prominent examples for the use of position in the very core of syntactical structure. English and Chinese both distinguish Agent and Patient by position only, German may do so as well:

English: (the)   man(Ag=Pos1)   hits (the)   enemy(Pt=Pos3)
Chinese: (Nàgè)   rén(Ag=Pos1)   jízhòngle   dírén(Pt=Pos3)
German: Paul(Ag=Pos1)   schlägt   Peter(Pt=Pos3)

Law in Language

Languages that do not use position must resort to designation, in other words they must have special formants designating at least one of the two semantic roles. German, Latin, Russian and many other languages designate both roles. In this case, Agent and Patient may, of course, change their positions:

German: der(Ag) Mann   schlägt den(Pt) Freund
         den(Pt) Freund schlägt der(Ag) Mann

As soon as an action involves three substance, position can no longer be used for all three of them. English as well as Chinese must specify at least one of the semantic roles by designation:

English: (The) man(Ag=Pos1) gives (the) ball(Pat=Pos3) **to** (his) friend

Chinese: nánzi(Ag=Pos1) bǎ   qiú(Pt=Pos3) jiāogěi péngyǒu
         man(Ag=Pos1)   take ball(Pt=Pos3) give   friend

While English resorts to designation ('to') in order to specify the person receiving the object (friend), Chinese cuts the action itself into two halves, so to speak. German designates all three semantic roles as do most Indo-Germanic languages. In this case position is no longer needed to specify roles - in other words, it may be freely changed - producing theoretically 3 times 2=6 alternatives for 'the man gives the ball to the friend'.

German: der(Ag) Mann   gibt dem(Rc) Freund den(Pt) Ball.
         der(Ag) Mann   gibt den(Pt) Ball   dem(Rc) Freund
         den(Pt) Ball   gibt der(Ag) Mann   dem(Rc) Freund
         den(Pt) Ball   gibt dem(Rc) Freund der(Ag) Mann*
         dem(Rc) Freund gibt der(Ag) Mann   den(Pt) Ball
         dem(Rc) Freund gibt den(Pt) Ball   der(Ag) Mann*

It is interesting to note that merely two of these six alternatives - though even these would still be perfectly understood - are hardly used - the ones where the agents occupies the last position. Such a positional change of the Agent is still possible with only two semantic roles (den(Pt) Freund schlägt der(Ag) Mann), but not with three. In Latin even this case would pose no problem, but the German language is already on the way of using position along with designation (Paul(Ag=Pos1) schlägt Peter(Pt=Pos3), so German is no longer completely free in positional choice.

Law in Language

The action of giving something to somebody else, though permitting a clear distinction of the agent and the receiving person, does not have a real patient. It is by mere formal extension that both English and Chinese make ball appear in the formal slot (=Pos3) reserved for true patients - as in 'he hits Paul(Pt=Pos3)'. It is for this reason that this pseudo-patient may sometimes change position:

a) John(Ag=Pos1) smeared (the) wall(Pt=Pos3) **with**(instr. des.)   paint
b) John(Ag=Pos1) smeared   paint(Pt=Pos3) **on**(loc. des.)   (the) wall

In the first case we have the instrumental designator 'with', in the second case the local designation 'on'.

When Agent and Patient are distinguished not by designation but by position, there seems to be no other alternative as the two positions *before and after the action*. Languages like Japanese that place the action at the end must resort to designation of *at least one* of the semantic roles:

English: (the)   man(Ag=Pos1)   hits (the)   enemy(Pt=Pos3)

Chinese: (Nàgè)   rén(Ag=Pos1)   jízhòngle   dírén(Pt=Pos3)

German: Paul(Ag=Pos1)   schlägt   Peter(Pt=Pos3)

Japanese: otoko   (wa)   Yuujin **wo**(=Pat)   utsu
               man   friend(Pt=Des)   hits

Now, it is an interesting problem to be dealt with by Constructional Linguistics whether we could expect any language that formally realizes Agent and Patient by position to change their respective place so that the former comes after and the latter before the action? I am unable to prove that this alternative would be impossible but the fact that so many actions do not have a patient (he sings, he runs, he thinks...) seems to make this alternative rather improbable. But, of course, this is not a proof.

I suspect that those whom in the preface I designated as the poets among linguists are hardly interested in possible combinations and laws but want to know what possibilities the choice of designation versus position offers or denies to the wealth of expression? I agree that this is a very important question indeed but it is beyond the scope of a grammar that wants to draw a line between what is arbitrary and what is lawful in language.

4 Fourth basic law (Pre- or postpositions)

It is a well-known fact that languages prefer pre- or postpositions according to where they posit verbs. Is this a purely haphazard phenomenon or one that may be explained by the constraints of formal realization? (In the following examples, verbs are underlined, pre- or postpositions appear in italics).

E:	(The) woman	travelled	*to*	(a) friend
E:	Noun	<u>Verb</u>	*Prep*	Noun

J:	Josei (wa)	yuujin	ni	ryokoushita
	(The) woman	friend	Postp	travelled
J:	Noun	Noun	*Postp*	<u>Verb</u>

The reason why Japanese could not possibly choose a word order with prepositions: J*: Noun *Prep* Noun <u>Verb</u>

cannot be derived from this simple example, but it becomes evident once we ask for the formal realization of a bound synthesis semantically specifying "friend" (specification = she had seen in Tokyo). Let's first turn to the English example:

E:	(The) woman	<u>travelled</u>	*to*	(the) friend she had seen	*in*	Tokyo
E:	Noun	<u>Verb</u>	*Prep*	Noun	<u>Verb</u>	*Prep*	Noun

J:	Josei (wa)	Toukyou	de	mita	yuujin	ni	ryokoushita
	(The) woman	Tokyo	in	had seen	friend	to	travelled)
J:	Noun	Noun	*Postp*	<u>Verb</u>	Noun	*Postp*	<u>Verb</u>

While there is no cogent reason why in the simple Japanese sentence 'Woman friend Postp <u>travelled</u>' the postposition could not be substituted with a preposition leading to 'Woman Prep friend <u>travelled</u>", such a formal realization (word order) would be quite impossible in the second example because then *two prepositions* would have to follow each other (note that Japanese has no relative pronouns for relative clauses).

J*: Woman *to* (=Prep) *in* (=Prep) Tokyo had seen husband travelled.
J*: Noun *Prep*   *Prep*   Noun Verb   Noun   Verb.

For the same reason, English could well use a postposition in our first simple example:

E*: (The) woman travelled (her) friend *to*
E*:   Noun Verb   Noun   *Postp*

But such a type of formal realization becomes impossible as soon as we semantically specify the object:

E*: (The) woman travelled (the) friend seen Tokyo *in*   *to*
E*:   Noun Verb   Noun Verb Tokyo *Postp Postp*

In other words, the constraints of formal realization explain why most languages where verbs precede objects will have prepositions while most languages where verbs follow them will instead make use of postpositions. There are, however, exceptions to the general rule in languages, *which are free to place verbs at different positions*, like Latin, Sanskrit and even German etc.

The preceding demonstration is an example of **constructive linguistics**. Once an initial choice as to the position of verbs has been made, what are the formal consequences? Are languages still free to choose between post- and prepositions? From examples such as these we may conclude that formal constraints become evident not in simple phrases but only when our analysis reaches to deeper levels. In the present instance, we have to ask how languages formally realize the bound synthesis!

5 Fifth basic law (concerning morphology)

Bound formants, that is all those that convey meaning only when occurring as the affixes of free formants, may only be used for realizing closed-field semantic categories like differences in number (singular/ plural, dual …), differences in time (present, past, future …) differences in person (I, you, he, she, it – combined or not with number, gender ...), differences in gender (he, she, it), differences in the shape of objects (round, long ...), differences in social position (high, low, neutral ...) and so on. This is a conditional law, some languages like Chinese have no bound formants.

Index

Action Synthesis
    monovalent 101
Action Synthesis, trivalent 54
Agent/ Patient 50
Aristotle 25
Bloomfield, Leonard 25
Boas, Franz 102
Bound formant 59
Categories of Semants 41
Chomsky, Noam 2, 32, 59, 126
Commands 63
Composites 103
Concepts, abstract 95
Conjunctions 41, **56**, 63, 95
    frozen 100
    logical 57, **60**
    observer 56
    time-space 56, 59
Connections 41, **56**
Constructive Linguistics 103
Copula 28
Dakota 102
Descriptive Linguistics 114
Desiderative 59
Differential Analysis 114
Differentiation Value 42
Distributionalism 25
Embedding 69
English 103

Eskimo 102, 103
Extension, formal 97
Formal Abbreviation 58
Formal Deficiency
    synchronous/ contextual 125
Formal Ellipsis 68, 89, 92
Formal Equivalence/ Abundance/
    Deficiency **107**, **114**, **117**
Formal Extension 50
Formal Suppression 117
Formant
    dummy 58, 65, 92
Formants 42
General Statements 96
Generativeness
    general 33
    speaker's 32
German 103
Grammar
    normative 32
    pedagogic 32
    scientific/pedagogic 31
Hallpike, Christopher 17, 23
Harris, Zellig S. 25
Humboldt, Wilhelm 25
Hymes, Dell 102
Infinitive 58
Informational Shifting (topic/ novum) 54,
    64, 65, 90

Informational Structure of Meaning 64
Intonation 42
Jackendorf, Ray 23
Japanese 53, 59, 91, 98
Jespersen, Otto 17, 47, 79
Kwakiutl 102
Logical colon 48
Logical head 48
Logical Head 95
Logical Structure of Meaning 59, 60, 63, 64
Lyons, John 59
Mahabharata 60
Meaning
   logical/ informational 63, 64
   superimposed 99
Mendivil-Giro, J. 14
Nootka 53
Novum (comment)/ Topic 54, 64
Panini 32
Paratactic Formal Classes 96
Passive Voice 66, 91, 98, 115, 117
Pinker, Steven 2, 14, 15, 16, 69, 79
Position 42
Possession Synthesis 54
Predicate 48
Progressive 53
Quality synthesis 100
Rank-lifting 94, 96, 97
Rank-Lifting 67, 120
Ranks
   semantic 47, 95, 101
Recursion 42, 69
Relative Clause 98
Relative Pronoun 72

Sanskrit 59, 60
Semantic Effacement 51, 58, **89**
   total 94
Semantic Inversion 41, 53, 65, 115
Semantic Roles 94
Semantic Tingeing 94, **99**
Semants 41, 45
   open field/ closed field 41
Sentence 42
Situational Context 68
Style 60
Subject 48
Subject/ Predicate 64
Substance
   indefinite 89
Syntheses
   enlarged 51, 101
Syntheses 41, 45
Syntheses
   free/ bound 98
Syntheses
   frozen 100
Tewa 102
Tones 42
Trendelenburg, F. A. 25
Types of conjunction **56**
Types of Synthesis **49**, 61
Unit of Information 71
Weinrich, Harald 25
Weisgerber, Leo 25
Whitney, W.D. 60
Whorf, Benjamin L. 25
Wittenberg, Eva 23
Zero-Form 68, 117

www.ingramcontent.com/pod-product-compliance
Lightning Source LLC
Chambersburg PA
CBHW021824170526
45157CB00007B/2679